CO-AYH-250

Greening the Wind

*Environmental and
Social Considerations for
Wind Power Development*

George C. Ledec
Kennan W. Rapp
Roberto G. Aiello

THE WORLD BANK
Washington, D.C.

Energy Sector Management Assistance Program (ESMAP) reports are published to communicate the results of ESMAP's work to the development community with the least possible delay. Some sources cited in this paper may be informal documents that are not readily available.

ISBN: 978-0-8213-8926-3
eISBN: 978-0-8213-8929-4
DOI: 10.1596/978-0-8213-8926-3

Library of Congress Cataloging-in-Publication Data has been requested.

Contents

Figures

Tables

Foreword

According to the International Energy Agency, electricity demand is projected to grow by around 30 percent by 2020. With growing attention to climate considerations and cost concerns regarding thermal generation as petroleum prices rise, alternative generation technologies will become increasingly important in tackling energy security issues. Renewable energy, and wind energy in particular, will be key in helping to meet this growing demand in a sustainable way.

As wind power is one of the most promising renewable resources in existence today, it offers several potential benefits. In the past decade, the use of wind power has expanded dramatically around the world, boosted by favorable economics, increasing interest in low-carbon technologies, and supportive governmental policies driven by the numerous benefits wind can provide. Since wind power does not rely on imported fuels, it helps countries to better use local resources while diversifying generation sources. Furthermore, by relying less on petroleum-based fuels with potentially high and volatile prices, wind power can help reduce the risk profile of the sector. In addition, if wind energy displaces fossil fuel–generated electricity, it can reduce carbon emissions in a cost-effective manner. Reducing carbon emissions has both environmental and economic value; the latter will become more evident as future uses of carbon-intensive fuels are further constrained. Wind power also has the potential to reduce air pollution at the local level by replacing more polluting sources of power generation, thereby improving environmental and human health. Finally, wind power stands out among all major power generation technologies (renewable or not) in that it requires almost no water, an increasingly scarce resource in much of the world.

However, while wind power development has many environmental and social benefits, it also poses various challenges, in particular for biodiversity and local communities. Therefore, development of wind power must be approached from multiple angles and advanced in an environmentally and socially sustainable manner. Specifically, wind power development has important implications for biodiversity, visual impacts, noise, radar and telecommunications infrastructure, access roads, land acquisition, and benefits-sharing, all of which should be considered in the wind power development process.

The World Bank is committed to supporting sensible paths toward sustainable development. As part of its efforts, the Bank recognizes the importance of global knowledge-sharing on important issues related to sustainable development. Building on the experience of the Sustainable Development Department of the Bank's Latin America and Caribbean Region with wind power projects in Latin America, *Greening the Wind: Environmental and Social Considerations for Wind Power Development in Latin America and Beyond* examines key topics for consideration during the wind power development process with an emphasis on the complex social and environmental challenges that may arise, particularly with large-scale, grid-connected onshore wind farms.

Wind power is an important part of the solution for the energy challenges that lie ahead. It is hoped that by sharing this report with industry, development organizations, research centers, governments, and NGOs, the power of the wind can be better exploited with less negative impact, therefore maximizing its benefits worldwide.

Philippe Benoit
Sector Manager, Energy Unit
Latin America and the Caribbean Region
World Bank

Acknowledgements

This report was prepared within the Energy Unit of the Sustainable Development Department of the Latin America and Caribbean Region of the World Bank. The Study Team was led by George C. Ledec and Roberto G. Aiello, with major inputs from Kennan W. Rapp and additional contributions from Pamela Sud, Almudena Mateos, and Megan Hansen. The report also benefitted from the suggestions and input from expert outside consultants, particularly Robert Livernash (social issues); Edward Arnett (bats); and Carl Thelander and Ed West (birds). The authors gratefully acknowledge the helpful contributions and comments of our World Bank Group reviewers, Soren Krohn, Lakhdeep S. Babra, Glenn Morgan, Juan D. Quintero, Haddy Sey, Richard Spencer, Francis Fragano, Jorge Villegas, and Dana Younger; as well as from other reviewers, Rolf-Guenter Gebhardt, Diane Ross-Leech, Wayne Walker, Carlos Gasco Travesedo, Rafael Villegas Patraca, Michael Fry, and Andrea Pomeroy. From outside the Bank, Doug Mason, Paula Posas, William Evans, and Robert Goodland also provided useful information to the Study Team. Thanks are owed as well to those who have worked on the three wind power projects that were selected as case study examples for the report. This includes Walter Vergara, Alejandro Deeb, Alonso Zarzar, Adriana Valencia, Irene Leino, and Karen Bazex of the World Bank; Carlos Sánchez Cornejo and Zirahuén Ortega of the Federal Electricity Commission (CFE), Mexico; and implementation staff from the Medellin Public Companies (EPM) in Colombia, and the National Administration of Power Plants and Electric Transmission (UTE) in Uruguay. Lynn Zablotsky helped to edit the final manuscript. Finally, we wish to thank our supportive managers in the World Bank's Latin America and Caribbean Region: Laura Tuck, Director; Philippe Benoit, Sector Manager; and Jocelyne Albert, Climate Change Coordinator, as well as the World Bank staff who helped to obtain the funding for our study. The financial and technical support by the Energy Sector Management Assistance Program (ESMAP) is gratefully acknowledged.

Acronyms and Abbreviations

APWRA	Altamont Pass Wind Resource Area
BFD	bird flight diverter
CDCF	Community Development Carbon Fund
CDM	Clean Development Mechanism
CER	Certified Emissions Reductions
CFE	Comisión Federal de Electricidad (the Mexican Government's Federal Electricity Commission)
CNH	critical natural habitat
dB(A)	A-weighted decibel(s)
EIA	environmental impact assessment
EIL	Enercon India Limited
EMP	Environmental Management Plan
EPM	Empresas Públicas de Medellín
GEF	Global Environment Facility
GHG	greenhouse gas/gases, including carbon dioxide and methane
GW	gigawatt(s)
GWh	gigawatt hours
ha	hectare(s)
ha/MW	hectares per megawatt
IBA	Important Bird Area
IBRD	International Bank for Reconstruction and Development
IFC	International Finance Corporation
INECOL	Instituto de Ecología (Ecology Institute, Mexico)
IPP	Indigenous Peoples Plan
km	kilometer(s)
kph	kilometers per hour
kW	kilowatt(s)
kWh	kilowatt hour(s)
LAC	Latin America and the Caribbean
m	meter(s)
m/sec	meters per second
mph	miles per hour
MVP	Monitoring and Verification Protocol
MW	megawatt(s)
NGO	nongovernmental organization
NIMBY	"not in my backyard"
PCR	physical cultural resources

ROW	right(s)-of-way
RPM	revolutions per minute
RSA	rotor-swept area
SA	social assessment
SEA	strategic environmental assessment
UK	United Kingdom
UN	United Nations
U.S. and USA	United States of America
UTE	Administración Nacional de Usinas y Transmisiones Eléctricas (Uruguay's national electricity company)
V	volt
WRA	wind resource area

Executive Summary

This report identifies good practices for managing the key environmental and social issues associated with wind power development and provides advice on how best to address these issues in project planning, construction, and operation and maintenance. It provides detailed background information on wind power, with special focus on two emerging themes of growing scientific and public interest: namely the biodiversity-related impacts and the broader socioeconomic and cultural dimensions of wind power development. Like wind power itself, the scope of this report is worldwide although special attention is paid to the issues characteristic of the Latin America and Caribbean (LAC) region. While the principal focus is on land-based wind power, it also briefly addresses the environmental and social impacts related to offshore wind development.

Overview

Wind power today is widely regarded as a key component of an environmentally sustainable, low-carbon energy future because it is renewable, requires almost no water, and generates near-zero emissions of greenhouse gases and other pollutants. In many parts of the world, wind power has the potential to significantly reduce greenhouse gas (GHG) emissions from electric power generation, thereby helping to limit the severe environmental and social consequences of human-induced climate change. The growth of wind power has also occurred due to its other positive attributes, including growing economic competitiveness. Wind power has become the world's fastest-growing source of power generation. While most wind power development to date has been in Europe, Asia, and North America, it is also poised to grow rapidly in other regions, including Latin America.

Notwithstanding its benefits, wind power poses several significant environmental and social challenges, most notably: (i) biodiversity-related impacts, (ii) visual and other local nuisance impacts, and (iii) a variety of socioeconomic and cultural concerns. To realize the full economic and climate change benefits of wind power, it is important to minimize the potential adverse environmental and social impacts of wind farms and their associated infrastructure, using measures such as those described in this report. Doing so will help to ensure adequate public acceptance of wind power and the fulfillment of its promise as an environmentally sustainable energy source.

Challenges

The adverse biodiversity-related impacts of wind power facilities mainly involve birds, bats, and natural habitats:

- **Birds** can be killed by collisions with wind turbines and sometimes also the guy wires of meteorological towers, at times in potentially significant numbers from a conservation standpoint. Bird species groups of special concern are birds of prey such as raptors, seabirds, migratory species, and grassland birds with aerial flight displays. Although modern large turbine blades appear to be moving slowly when viewed from a distance, the blade tip speed is actually quite fast (up to about 270 kilometers per hour [kph]), such that the birds are struck by surprise. However, for some scarce, open-country species such as prairie grouse, the main conservation threat posed by wind power development is not collisions, but rather displacement from their habitat because the birds instinctively stay far away from wind turbines, transmission towers, and other tall structures.

- **Bats** tend to be killed by wind turbines at higher rates than birds, in part because bats are apparently attracted to wind turbines. As bats are long lived and have low reproductive rates, they tend to be more vulnerable to the added mortality from wind turbines than most faster-reproducing small bird species.

- **Natural habitats** can be lost or fragmented when they are cleared to establish wind power facilities, sometimes with significant risks to biodiversity. For example, wooded mountain ridge-tops, particularly in the tropics, often harbor unique plant and animal species, due in part to their wind-swept micro-climate. Long rows of turbines with interconnecting roads along such ridge-tops can disproportionately affect scarce, highly localized species. Constructing access roads to previously remote wind farm sites can also lead to the loss or degradation of natural habitats, either directly, through road construction and resulting erosion, or indirectly, through increased land clearing, wood cutting, informal mining, hunting, or other human activities facilitated by improved access.

Wind power development involves local nuisance impacts that are sometimes of considerable public concern. Foremost among these are the visual impacts of large wind turbines and associated transmission lines, which some people regard as an eyesore. Visual impacts can be particularly sensitive in areas with high tourism potential and in popular recreation areas. To date, visual impacts have emerged as a leading socio-environmental constraint to installing new wind farms and associated transmission lines. This concern has been most pronounced in North America and Europe, but is also becoming evident in some developing countries. A special type of visual impact, shadow flicker, can be a problem when turbines are located relatively close to homes, creating an annoying effect of rapidly blinking shadows when the sun is near the horizon. Another type of nuisance impact is noise. Noise from wind turbines is evident only at rather close range (within 300 meters [m]), even though it is often mentioned as a concern by local residents when a new wind farm is proposed. In addition, electromagnetic interference represents yet another type of nuisance impact as it can limit the functioning of aviation radar, radio, television, and microwave transmission systems when operating wind turbines are within the line-of-sight of the radar or telecommunications facility. Furthermore, wind turbines can pose an aviation safety risk when they are located too close to airports or in agricultural areas where aerial spraying of pesticides takes place. Finally, there can also be very slight public safety risks from blade or ice throw.

Wind power development entails a variety of socioeconomic and cultural issues that need to be carefully addressed. Regarding the maintenance and/or enhancement of people's livelihoods, the impacts of a typical wind farm project are often positive. For instance, the nature of the footprint of wind power projects permits most preexisting land uses to continue. When direct economic benefits are factored in, such projects can increase income for rural landholders in the wind farm area and help generate employment and boost local economies. Land acquisition is typically done through negotiation of lease/rental or royalty payments, or less commonly, through outright purchase or expropriation. Whatever method is used, acquiring land without effective measures to address adverse impacts such as the full replacement of lost assets can lead to protracted negotiations between project sponsors and affected landholders over compensation terms. Under certain circumstances it can also lead to more serious social conflict.

Many of the areas with high wind power potential such as remote deserts, plains, and mountaintops are where indigenous peoples and other traditional rural populations tend to be found. Under such circumstances, the introduction of a wind farm by an external developer can unleash forces of cultural change that those living in the project influence area might find undesirable or even harmful. In addition, without adequate planning and care during project construction, wind projects can also sometimes damage physical cultural resources (PCR).

Managing Local Environmental and Social Aspects

Typically the single most important measure for managing environmental and social impacts is careful site selection of wind power facilities. Since many countries' potential wind resources remain largely untapped, they typically have multiple options regarding where to locate new wind farms, for connection to a national or regional electricity grid. From a biodiversity standpoint, the lower-risk sites for wind power development tend to have low bird and bat numbers year-round and do not harbor species or ecosystems of conservation concern. In fact, careful site selection is the most important measure to avoid or minimize other kinds of adverse local impacts, including visual impacts, noise, and electromagnetic interference. Careful site selection is also particularly important when considering wind power development on indigenous community lands. In addition, once the general site for a new wind farm has been selected, some negative impacts can be avoided or further reduced by adjusting the location of turbine rows and even individual turbines. The selection of wind power equipment, taking into account turbine size and other specifications, can also influence biodiversity as well as visual impacts.

A variety of planning tools are available to optimize site selection and manage environmental and social impacts. These tools include:

- **Strategic environmental assessments** (SEAs), a key tool for site selection of wind power facilities, typically produce overlay maps that show where the zones of high wind power potential—based mainly on wind speeds and proximity to the power transmission grid—are located in relation to the areas of major environmental and social sensitivity. Some SEAs also generate zoning maps for prospective wind power development. Specifically, SEAs may recommend where wind farms or transmission lines should be prohibited, allowed only with special precautions, or actively promoted. SEAs are also important in terms of assessing the cumulative environmental impacts of multiple wind farms within a wind resource area (WRA).
- **Environmental impact assessment** (EIA) reports represent an essential tool for identifying and managing environmental impacts at the project level. An EIA report for a wind project is most useful when it includes an environmental management plan (EMP) that specifies each of the actions to be taken during project construction and operation to mitigate any adverse impacts and enhance any

positive ones. Although many EIA studies cover social impacts (along with the biological and physical ones), it may be helpful to also carry out separate social assessments (SAs), particularly for those wind power projects that could potentially affect (either positively or negatively) indigenous peoples or other vulnerable groups.

■ **Project legal agreements** need to reflect agreed wind farm operating standards, which might specify post-construction monitoring and data-sharing, operational curtailment, and equipment and landscape maintenance. In this regard, environmental and social mitigation measures are much more likely to be implemented if they have been explicitly described and budgeted for in signed project agreements, bidding documents, and contracts.

Effective management of wind power impacts typically involves systematic stakeholder engagement. A key element of public consultation in the context of energy development is educating the public and decision-makers about the full range of trade-offs, impacts, and benefits associated with different technologies, including wind. While many people like wind power in general, they are often opposed to having wind farms or transmission lines in their "backyards." Therefore, a participatory approach regarding the location of wind project infrastructure can have a decidedly positive impact on public attitudes towards particular projects. In addition, the early dissemination of information on the implications of a proposed wind power development project—for both the natural environment and local people—can facilitate thorough consideration of all relevant trade-offs during the decision-making process.

People who own or use land required for a wind project need to be suitably compensated. Land acquisition for wind power facilities frequently takes place through lease or rental payments, which often allow certain preexisting land uses to continue. Sometimes land ownership is transferred to the wind project owner or operator, through either voluntary purchase or involuntary expropriation. When expropriation is used, providing for compensation to affected landholders at replacement cost, rather than relying on the cadastral values of the needed lands, is more satisfactory for the landholders as it enables the lost lands and related assets to be fully replaced based on market prices. Additional compensation payments might also be needed for other adverse impacts stemming from wind projects, such as damage to property or disruption of productive activities during project construction.

As with other types of large-scale energy development, equitable benefits-sharing for local residents is an important issue for wind project planners. Benefits-sharing is additional to the payment of compensation for lost assets and can take many forms, including: (i) payment of rents or royalties to affected landholders and neighbors, (ii) clarification of property rights for host communities during project preparation, (iii) employment opportunities for local workers during construction or operation of wind power facilities, (iv) local benefits programs, and even (v) community ownership of wind farms. A broad conceptualization of benefits is the recommended approach to address issues of fairness that are of concern to those whose lands are directly or indirectly affected by wind power development. By adopting a benefits-sharing approach that goes beyond those who are immediately and adversely affected by wind farm construction and operation, project sponsors can find themselves in a much better position to influence public perceptions, improve community relations, and engage in effective risk management.

During the installation of wind turbines, access roads, and transmission lines, the use of good construction practices will serve to minimize adverse environmental and social impacts. As with any large-scale civil works, environmental rules for contractors should be specified in bidding documents and contracts. In addition, adequate field supervision by qualified personnel, along with transparent penalties for noncompliance, is also needed.

Post-construction monitoring of bird and bat mortality is an indispensable tool for the environmental management of wind power projects. Such monitoring is needed to: (i) determine whether or not a significant bird or bat mortality problem exists at a given wind farm, (ii) predict the biodiversity-related impacts of scaling up development within a particular wind resource area, (iii) enable adaptive management of wind farm operation to reduce bird or bat mortality, and (iv) advance scientific knowledge in a field that still faces a steep learning curve. Post-construction monitoring is carried out during the first two years or so of wind farm operation, and continued if significant mortality is found so that mitigation measures can be tested and implemented. The data collected from each wind project should ideally be presented in a readily understood form, publicly disclosed, and collaboratively shared with other wind developers, regulatory agencies, and international scientific research networks and partnerships. Regarding bird and bat monitoring, it is also important to note that there is a very real—and sometimes significant—difference between real mortality and observed mortality at wind farms; therefore, appropriate correction factors need to be used.

Certain changes in wind turbine operation can lead to substantial reductions in bat or bird mortality. For bats, the most important turbine operation change appears to be an increase in cut-in speed, which is the lowest wind speed at which the rotor blades are spinning and generating electricity for the grid. For migratory birds, the most important operating change is often short-term shutdowns, in which the rotor blades do not turn during peak migration events. Wind project planners can calculate the extent to which these types of operational curtailment would likely affect power generation and financial returns, and compare them with the anticipated reduction in bat or bird mortality.

Wind farm maintenance practices can be an important tool for managing environmental and social impacts. Diligent equipment maintenance—such as capping holes in wind turbine nacelles—can help prevent unnecessary bird mortality or other environmental damage. In addition, landscape management at wind farms needs to consider a variety of environmental and social objectives. For best results, vegetation management at wind farms is carefully planned in advance, discussed with stakeholders, and recorded within the project's EMP. Furthermore, managing public access to wind power facilities needs to take into account a variety of different environmental and social objectives. Conservation offsets can also be useful in mitigating biodiversity impacts from wind projects and enhancing the projects' overall conservation outcomes.

Conclusions

Table 5.1 summarizes the main environmental and social impacts typically associated with wind power projects. The table sets out the options available for the mitigation of the negative impacts and enhancement of the positive ones of wind power projects, specifically in terms of: (i) biodiversity impacts, (ii) local nuisance impacts, and (iii) socioeconomic and cultural impacts.

Wind power is an important part of global efforts to meet increasing energy needs. As many countries rapidly scale up their wind power development, ensuring that environmental and social impacts are adequately addressed will enhance the sustainable development benefits of this renewable energy technology.

CHAPTER 1

Objectives and Scope
of this Report

Overview

This report identifies good practices for managing the key environmental and social issues associated with wind power development and provides advice on how best to address these issues in project planning, construction, and operation and maintenance. It provides insights and good practice suggestions for managing the environmental and social aspects of wind power projects, along with the analysis, case studies, scientific evidence, and references upon which this advice is based. Besides providing an overview of wind power development and the accompanying environmental and social issues, this report focuses on two emerging themes of growing scientific and public interest: (i) biodiversity-related impacts and (ii) the broader socioeconomic and cultural dimensions of wind power development.

Complementarity to Existing World Bank Group Guidelines

This report aims to complement, but not replace, the 2007 *Environmental, Health, and Safety (EHS) Guidelines for Wind Energy*, which form part of the World Bank Group's *Environmental, Health, and Safety Guidelines* for all types of development projects (superseding the earlier *Pollution Prevention and Abatement Handbook*). The *EHS Guidelines* provide a systematic overview of the typical environmental, as well as occupational health and safety, issues associated with wind farms. In this regard, this report provides additional guidance on how to address specific complex environmental and social issues that have recently emerged as significant concerns in wind power projects, such as bird and bat conservation, visual impacts, and community benefits-sharing arrangements. The advice contained in this report does not constitute World Bank policy or guidelines. Rather, it is hoped that wind project planners and developers will make use of both documents as needed.

Relevance to World Bank Safeguard Policies and IFC Performance Standards

This report does not establish any new environmental or social policy or guidelines for the World Bank Group. Instead, it offers technical advice aimed at strengthening the design and operation of wind power projects in a manner consistent with existing World Bank Safeguard Policies and International Finance Corporation (IFC) Performance Standards.[1]

Scope

Like wind power itself, the scope of this report is worldwide, although special attention is paid to the issues characteristic of the Latin America and Caribbean (LAC) region. Accordingly, in addition to drawing from the experiences of wind projects around the world, the discussion relies significantly on the lessons learned from three case studies involving World Bank–supported wind power projects in Mexico, Colombia, and Uruguay (Appendixes A–C). While the focus is mainly on the environmental and social impacts associated with large-scale, grid-connected wind farms, some of the advice provided may also be applicable to small-scale, off-grid wind turbines. In addition, the impacts associated with power transmission lines and access roads are addressed as well since these works are an essential component or precondition of any grid-connected wind power project—as is the case for most other types of power generation. Finally, while the report's principal emphasis is on land-based (onshore) wind power because, for economic reasons, it is still heavily preferred within LAC and other developing regions, the environmental and social impacts related to offshore wind development are also briefly discussed.

Intended Audience

This report is intended for anyone with a strong interest in wind power and its environmental and social implications, including staff of the World Bank Group and other international development organizations; wind project investors and operators; government officials; energy and power sector planners; nongovernmental organizations (NGOs) with an environmental, social, energy, and/or scientific focus; news media and bloggers; and any interested members of the public. Many of the environmental and social issues and approaches discussed in this report are likely to be relevant for most wind power projects, even offshore ones to some extent.

Notes

1. The World Bank Safeguard Policies that are most likely to apply to wind power projects are: (i) Environmental Assessment (OP/BP 4.01), in all cases involving World Bank–supported investment projects; (ii) Involuntary Resettlement (OP/BP 4.12),when compulsory land acquisition is required for the construction of the wind farm and associated infrastructure; (iii) Indigenous Peoples (OP/BP 4.10) when the wind project involves indigenous communities; (iv) Natural Habitats (OP/BP 4.04), when natural land areas (including the biologically active airspace above them) are significantly affected; (v) Forests (OP/BP 4.36), when forests of any kind would be affected; and (vi) Physical Cultural Resources (OP/BP 4.11), when wind project construction might affect archaeological, historical, or sacred sites or objects. Also, OP 8.60 on Development Policy Lending applies to World Bank support of specific country policies—potentially including wind power promotion—that are likely to "cause significant effects on the country's environment, forests, and other natural resources." The IFC Performance Standards that are especially applicable to wind projects include (i) Performance Standard (PS) 1, Social and Environmental Assessment and Management System; (ii) PS 2, Labor; (iii) PS 4, Community Health and Safety; (iv) PS 5, Land Acquisition and Involuntary Resettlement; (v) PS 6, Biodiversity Conservation and Sustainable Natural Resource Management; (vi) PS 7, Indigenous Peoples; and (vii) PS 8, Cultural Heritage.

CHAPTER 2

Overview of Wind Power Development

Introduction

As mentioned in Chapter 1, this report identifies good practices for enhancing the environmental and social outcomes of wind power development, at the level of individual projects and for broader-scale planning and policy formulation. While humankind has harnessed wind power for centuries on a small scale for sailing ships, water pumps, and other uses, the large-scale generation of electric power with wind turbines is a relatively recent development. Wind power has quickly emerged as a highly promising, economically feasible, fully renewable, and very low-carbon technology for generating electricity. Nonetheless, the rapid expansion of wind power development comes with its own set of particular environmental and social sustainability issues, even as it helps to mitigate the enormous environmental and social risks posed by human-induced global climate change.

Rapid Growth of Wind Power

Over the last 20 years, wind power has become the world's fastest-growing source of power generation. Globally, wind power installed capacity has increased nearly 10-fold during the past decade (Figure 2.1). In 2010 alone, about 38,000 MW (megawatts) of new wind generation capacity were installed, with a worldwide investment of about US$55 billion[1] and bringing the world total to 196,630 MW (World Wind Energy Association 2010). This represents a one-year increase in global installed capacity of about 23 percent.

The World Wind Energy Association notes that in 2010, China and the United Statestogether represented 43 percent of global installed wind power capacity. The top five countries (China, United States, Germany, Spain, and India) represented 74 percent of worldwide wind capacity in 2010. At least 83 countries are today using wind power on a commercial basis, and 35 countries have wind farms with an installed capacity exceeding 100 MW. In 2010, wind turbines collectively generated an equivalent of about 2.5 percent of global electricity worldwide. China, followed by the United States, is now the world's leader in installed wind capacity (Figure 2.2).

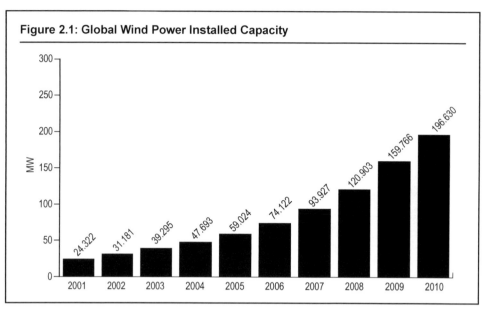

Figure 2.1: Global Wind Power Installed Capacity

Source: World Wind Energy Report 2010.

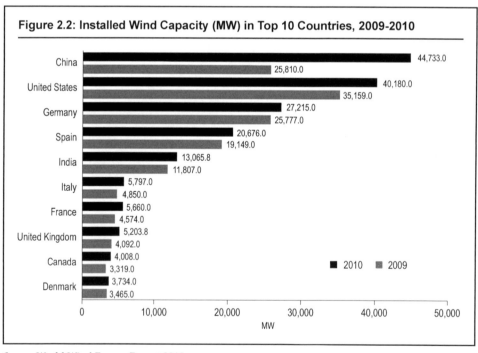

Figure 2.2: Installed Wind Capacity (MW) in Top 10 Countries, 2009-2010

Source: World Wind Energy Report 2010.

Wind Power in Latin America and the Caribbean

By the end of 2010, overall installed wind power capacity in the Latin America and Caribbean (LAC) region (1,983 MW) accounted for only 1 percent of global capacity. Currently, Brazil is in a position to establish itself as the leading wind country in the region given its strong domestic manufacturing industry, with several international companies already producing wind turbines in the country. Besides Brazil, new wind farms have been recently installed in Argentina, Chile, Costa Rica, Cuba, Mexico, Uruguay, and Jamaica. However, with the exception of Brazil and Mexico, most LAC wind markets can still be regarded as in a state of infancy. Some Latin American countries have very substantial, mostly untapped wind power potential. Interest in wind power in the region is growing, driven by a desire to diversify energy sources, concerns about high fossil fuel prices, and the availability of financial incentives such as carbon credits. Some countries, including within Central America, view wind power as a potentially useful, seasonal complement to hydropower since winds tend to blow more strongly during the dry season, while water for power generation is most abundant during the rainy season. As they scale up wind power development, LAC countries face the challenges of effectively managing biodiversity and other environmental concerns—in part by learning from mistakes made in some developed countries—as well as ensuring that community benefits-sharing and other social concerns are addressed both efficiently and equitably.

Prospects for Future Growth

In the foreseeable future wind power is poised for continued rapid growth worldwide. The exploitable global wind resource is large and reasonably well distributed across the five continents, though with much variation among countries. Modern wind turbines are series produced by a well-established industry, and can be installed fairly quickly. In terms of growth, significant near-term growth is expected in the leading wind markets of China, India, Europe, and North America. High growth rates are also projected in several LAC as well as new Asian and Eastern European markets. And in the mid-term, major projects are expected to be implemented in some African countries, notably in South Africa (with its feed-in tariff) and within North Africa.

Wind Power as "Green Energy"

The growing attractiveness of wind power to many governments, electric utilities, private investors, and the general public is due to several highly attractive features of the technology. Unlike fossil fuels, wind energy is an inexhaustible natural resource; most sites of high potential for wind power generation are likely to remain that way, even under typical scenarios of future climate change (IPCC 2007). Moreover, the generation of electricity from wind does not emit air or water pollution; relatively small carbon and other emissions are normally associated with the manufacture, transport, installation, and maintenance of wind turbines and associated infrastructure.[2] Furthermore, where it substitutes for coal- or oil-fired electricity generation, wind power can also reduce local air pollution problems. Finally, in contrast to most other power generation technologies (renewable or otherwise), wind power does not require any water in the production process, thus making it particularly suitable in water-scarce regions.

Mitigating Climate Change

There is an urgent need to expand the use of low-carbon power generation technologies due to the serious threats posed by greenhouse gas (GHG) emissions. The global climate change problem, along with the clear linkages to human-induced GHG emissions, is well described in numerous scientific (for example, IPCC 2007) and popular publications, and summarized in the Latin American context by de la Torre et al. 2009. While the most ambitious worldwide scaling up of wind power development will not adequately mitigate global climate change by itself; it stands to play a significant role, in combination with a wide range of other measures aimed at reducing and offsetting greenhouse gas emissions.

Environmental and Social Challenges

Over much of the world, wind power has come to be considered as a key component of an environmentally sustainable, low-carbon energy future. Nonetheless, wind power poses its own particular set of environmental and social challenges. Of special concern are the impacts on biodiversity (birds, bats, and natural habitats); visual and other local nuisance impacts; and a variety of socioeconomic and cultural issues, including stakeholder acceptability, compensation for affected assets, and benefits-sharing opportunities. As many countries rapidly scale up their wind power development, these environmental and social concerns will need to be properly addressed, both to ensure adequate public acceptance and to fulfill the promise of wind as an environmentally sustainable energy source.

Offshore Wind Power Development

This report's main focus is on land-based (onshore), rather than offshore, wind power generation because wind power investments to date in most developing countries worldwide, including Latin America, have been almost exclusively onshore. By contrast, offshore wind power investments are heavily concentrated in northern Europe, and there are plans for their rapid expansion in the United States. In addition, China is undertaking wind power development within coastal intertidal zones. By the end of 2010, total offshore installed capacity amounted to almost 3.1 gigawatts (GW), or 1.6 percent of total wind capacity worldwide. Figure 2.3 shows the top five countries in offshore wind power.[3]

Future Potential and Advantages of Offshore Wind Power Development

Offshore wind power development has tremendous future potential in much of the world along with certain key advantages in comparison with land-based wind power. First, offshore winds tend to blow more strongly and consistently, with lower turbu-

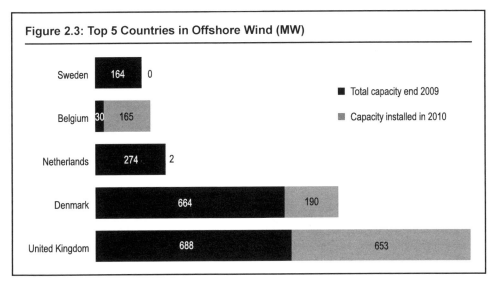

Figure 2.3: Top 5 Countries in Offshore Wind (MW)

Source: World Wind Energy Report 2010.

lence than on land. Second, turbines for offshore production can be built larger than the ones installed on land because very large parts such as rotor blades do not have to fit on roads of limited width, or under highway bridges, to reach wind farm sites. Nor do access roads need to be built or widened to reach offshore wind farms. Finally, since offshore wind power sites are often closer to electricity demand centers (for example, large coastal cities) than are many high-potential sites for land-based wind power, shorter transmission lines on land are typically needed. Nonetheless, offshore wind projects are more expensive to build and operate than those on land, with the cost often being up to double that of land-based projects (R. Gebhardt, *pers. comm.* 2010). Offshore wind power thus requires considerably greater capital outlays per MW installed. In this regard, since most countries have only begun to develop their onshore wind power potential, economic considerations still favor land-based over offshore development.

Environmental and Social Impacts of Power Generation Technologies

To put wind power into perspective, it is important to recognize that all large-scale, commercially available electric power generation technologies today come with their own set of significant environmental and social concerns. Table 2.1 provides a brief summary of the adverse environmental and social impacts associated with power generation from coal, petroleum, natural gas, nuclear, hydroelectric, solar, geothermal, biomass, and wind technologies. The optimum mix of power generation technologies in environmental, social, technical, economic, cultural, and political terms varies by country and sub-region although wind power is likely to be a desirable option in many areas. For those cases where wind power will be the generation technology of choice, this report outlines the good practices and policy options which, if applied, would help to maximize net environmental and social benefits.

Table 2.1: No Free Lunch—Environmental and Social Impacts of Power Generation Technologies

*This table does **not** seek to rank the power generation technologies by the importance of their impacts or by overall desirability. Nor is it a comprehensive listing of all significant environmental or related social impacts. Rather, its objective is to illustrate that all modern, commercial-scale power generation technologies involve important environmental and social impacts that need to be carefully considered in choosing the optimum mix of power generation technologies for a given country or region, and that these impacts need to be adequately mitigated to the extent feasible.*

TECHNOLOGY	MAJOR IMPACTS AND ISSUES
Coal	Largest GHG emissions per unit of useful energy delivered of any fossil fuel; severe air pollution and resulting health damage and acid rain; management of toxic waste ash from combustion; extensive and largely irreversible damage to land and natural habitat loss from open-pit and "mountaintop removal" coal mines; health and safety risks for workers in underground mines.
Petroleum (fuel oil, diesel)	Major GHG emissions from combustion, and (if natural gas is flared or vented) also from extraction; air pollution from combustion; oil spills from extraction and transport pollute ocean and inland waters, killing aquatic and marine life; wildlife mortality at waste oil ponds and from night-time gas flaring; transport and on-shore extraction can involve new roads or port facilities, sometimes in remote natural habitats and areas that are home to indigenous peoples or vulnerable minorities.
Natural Gas	Significant GHG emissions from combustion (also methane leakages), though proportionately less than coal or petroleum; transport and on-shore extraction can involve new roads or port facilities, sometimes in remote natural habitats and areas that are home to indigenous peoples or vulnerable minorities.
Nuclear	Risk of infrequent but potentially severe operating accidents; long-term storage needed for radioactive waste; weapons proliferation and terrorism risks associated with some nuclear power technologies; some GHG emissions from uranium mining and processing; varied environmental impacts from uranium mining.
Hydroelectric	Permanent flooding of land by reservoirs, sometimes resulting in large-scale human resettlement and/or major loss of terrestrial natural habitats; downriver hydrological changes; water quality deterioration can occur, upstream or downstream of the dam; loss of fish (especially migratory species) and other aquatic life due to changed river conditions, although some fish species might benefit; spread of floating aquatic weeds, mosquitoes, and water-related diseases in tropical zones; inundation of physical cultural resources; significant GHG emissions from those reservoirs that flood extensive forested areas; new access roads and overhead power lines often pass through natural habitats; some risk of causing earthquakes (induced seismicity).
Biomass (wood, crop residues, dedicated crops, and so forth)	Environmental impacts vary, according to how the feedstock is cultivated or harvested. Expansion of crop cultivation or non-native forest plantations can eliminate natural habitats, harming biodiversity and possibly resulting in significant net GHG emissions. Depending on the management practices used, harvesting of natural forests or other native vegetation might be environmentally sustainable (or not). Burning of crop residues for power generation can be environmentally benign, unless those same residues are needed by local populations for fuel, fertilizer, or livestock feed. Without adequate equipment, biomass combustion for power generation can cause air pollution.
Solar (photovoltaic and solar thermal)	Large land development footprint (around 3-4 ha/MW) that is devoid of plant or animal life (unlike the land flooded by hydroelectric reservoirs or within wind farms); some solar thermal technologies have significant water requirements (often a limiting factor in desert areas); new access roads and overhead power lines often pass through natural habitats. Distributed (for example, roof-top) solar generation is more environmentally and socially benign.
Geothermal	Land development footprint relatively small (similar to that of onshore oil or natural gas), but may be in remote, wild areas; produced toxic brine water requires careful management (ideally, reinjection); new access roads and overhead power lines often pass through natural habitats; some risk of causing earthquakes (induced seismicity).
Wind (land-based)	Bird and bat mortality at wind turbines (potentially significant for some species); visual impacts are sometimes highly controversial; wind farms, along with access roads and overhead power lines, often fragment natural habitats and make them unsuitable for some wildlife species; areas with high wind development potential include some plains and deserts that are inhabited by indigenous peoples or vulnerable minorities.

World Bank Group Support for Wind Power

To date, the World Bank Group has supported about 20 wind power projects, totaling about 950 MW of generating capacity and representing some US$367 million in investment. These projects have been funded through varying combinations of International Bank for Reconstruction and Development (IBRD) loans, International Finance Corporation (IFC) investments, Global Environment Facility (GEF) grants, carbon offset credits, government counterpart funding, and private sector financing. Presently, wind power projects with ongoing or anticipated Bank Group support are under advanced planning or implementation in every developing region, including Armenia, Bulgaria, Cape Verde, China, Chile, Colombia, the Czech Republic, the Arab Republic of Egypt, India, Jordan, Mexico, the Philippines, Poland, Sri Lanka, Turkey, and Uruguay (see Table 2.2).

Table 2.2: Wind Power Projects Financed by the World Bank Group

FY	Country	Project Name	Wind financing (million US$)	MW in operation
FY93	India	Renewable Resources Development	87.20	85.0
FY96	Sri Lanka	Energy Services Delivery	3.80	3.0
FY99	Cape Verde	Energy/Water Sector Reform	2.70	3.0
FY99	China	Renewable Energy Development	13.00	21.0
FY03	Colombia	Jepirachi Carbon Off Set	2.12	19.5
FY05	China	China Renewable Energy Scale-Up Program	67.00	100.0
FY05	Philippines	PCF-Northwind Bangui Bay	1.51	33.0
FY06	India	Enercon	8.00	75.0
FY06	Armenia	Renewable Energy	3.00	40.0
FY07	India	MSPL	33.00	36.6
FY07	China	Huitengxile Wind Farm	67.00	100.0
FY07	Mexico	Wind Umbrella (La Venta II)	20.53	83.3
FY07	Poland	Puck Wind Farm	2.81	22.0
FY08	Bulgaria	Bulgaria Wind Farm-AES Kavarna	40.00	156.0
FY08	Jordan	Promotion of a Wind Power Market	6.00	65.0
FY09	Chile	Norvind	70.00	46.0
FY09	Czech Republic	Various small wind projects	27.76[a]	18.5[b]
FY09	India	GPEC Wind Power	40.00	132.8
FY09	Uruguay	Uruguay Wind Farm	0.80	10.0
FY09	Turkey	Rotor Eleltrik	76.83	135

Source: Authors.
a. Total financing offered in the form of a partial loan guarantee for several small wind projects combined.
b. Total MW for several small wind projects combined.

Aside from direct investment, the Bank is playing a key role in creating a more conducive environment for wind power in developing countries through a combination of development policy lending, technical assistance, analytical studies involving power sector reforms and climate change mitigation, and policy dialogue. These mechanisms and activities are aimed at helping client countries overcome existing barriers to increased wind power development. Such barriers include inadequate tariff structures, lack of institutional capacity and information on wind resources, and lack of knowledge, particularly in the financial sector, about wind power and how to evaluate the risks involved in lending for it. In addition to the countries where wind power has received World Bank Group financing (Table 2.2), other countries where the Bank Group has provided technical assistance to facilitate wind power development include Argentina, Bolivia, Bosnia-Herzegovina, Djibouti, Egypt, Ethiopia, Ghana, Honduras, Kenya, Morocco, Mongolia, Pakistan, Peru, Slovakia, Syria, Tanzania, Timor Leste, and Yemen.

Wind Power Infrastructure

Wind power infrastructure includes principally: (i) the wind turbines themselves, usually grouped as a "wind farm"; (ii) the power transmission lines (usually overhead, outside the wind farm itself) and transmission substations; and (iii) the access roads needed for construc-

Figure 2.4: Typical Structural Components of a Wind Turbine

Source: World Bank Group 2007 Environmental Health and Safety Guidelines.

tion and maintenance of the wind farms and transmission lines. There are also certain ancillary facilities, such as substations, offices and control rooms, maintenance depots, storage sheds, parking lots, and, in more remote areas, construction camps. Figure 2.4 shows the structural portions of a typical wind turbine, while Figure 2.5 illustrates the main elements of a typical onshore wind farm.

Wind Power Footprint

The land footprint of an onshore wind farm is measured in several different ways:

- The **land area cleared** includes the space occupied by wind turbine platforms, access roads, parking lots, project offices, and other civil works, as well as any additional land cleared for the staging and maneuvering of heavy equipment used during turbine installation. The total land area cleared is variable, but is often in the vicinity of 1–2 ha/MW (hectares per MW). On a per-MW basis, wind farms comprised of larger turbines tend to require proportionately less land because the total number of turbines and turbine platforms, interconnecting road area, and related facilities is less.
- The **land area claimed** by a wind farm is the space within which the turbine array, substation, and other wind farm facilities will fit. There is also sometimes

Figure 2.5: Typical Elements of an Onshore Wind Farm

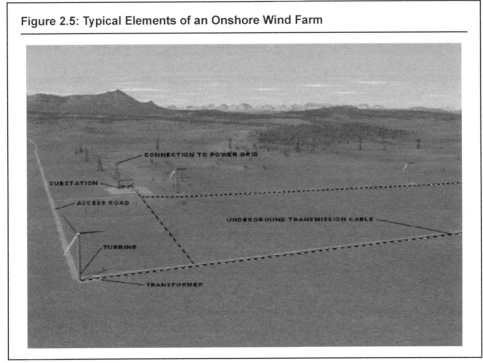

Source: World Bank Group 2007 Environmental Health and Safety Guidelines.

Photo: Roberto G. Aiello
The footprint of land cleared for wind turbine installation includes the area required for the staging of heavy equipment and large turbine parts, as shown here during wind farm construction in Uruguay.

an additional buffer area around the wind farm where landowners are prohibited from planting tall trees or erecting tall structures, designed to avoid affecting wind flows to the turbines. The wind farm area varies considerably based on wind conditions, topography, and other factors, but is often in the range of 10-30 ha/MW (sometimes more). Within the wind farm area, the roughly 90 percent of the land that is not cleared for wind power equipment is generally available to continue preexisting uses such as grazing or cultivation.

■ The **viewshed** or "visual footprint" of a wind farm is the area within which the wind farm is visible. This area varies according to topography and turbine heights, and can involve a radius of up to 30 km (kilometers) although the more significant visual impacts occur within about 5 km.

Notes

1. Assuming 1.3 million euros/MW installed, and 1 euro=1.3 US$.
2. The carbon footprint of wind power generation increases marginally when wind turbines are installed within forested areas such as many wind farms in the eastern US and some proposed ones in Panama because: (i) greenhouse gas emissions occur from the forest clearing needed to install wind turbines and access roads (typically around 1-2 ha/MW) and (ii) this deforested land no longer functions as a carbon sink.
3. By the end of 2009, wind farms installed offshore could be found in 12 countries, 10 of them in Europe and some minor installations in China and Japan.

CHAPTER 3

Making Wind Power Safer for Biodiversity

This chapter summarizes current knowledge about the biodiversity-related impacts of wind power in Latin America and worldwide. Despite its enormous advantages of renewability and near-zero carbon emissions, wind power can pose significant biodiversity-related environmental challenges. The discussion focuses mainly on land-based (onshore) wind power development as for economic reasons, it is still heavily preferred within Latin America and the Caribbean (LAC) and other developing regions. However, the section "Biodiversity Impacts of Offshore Wind Development" briefly addresses the biodiversity aspects of offshore wind power. In addition, specific attention is paid to large-scale, grid-connected wind farms although some advice may also be applicable to smaller-scale, off-grid wind power installations.

Wind power development includes: (i) the wind turbines themselves, usually grouped as a wind turbine facility or "wind farm," (ii) the power transmission lines and transmission substations, and (iii) the access roads needed for construction and maintenance of the wind farms and transmission lines. Scaling up wind power development often implies large-scale expansion of power transmission grids for several reasons. First, the highest-potential land-based wind resource areas (WRAs)—those with relatively steady, high winds—are often a long distance from the urban centers of electricity demand. Second, wind power (the availability of which at any given site varies daily and seasonally) needs to be complemented with other power generation sources, which need to be interconnected.

Biodiversity Impacts of Wind Power

Overview

There is considerable scientific consensus that human-induced global climate change poses serious, even catastrophic, risks to the long-term survival of many animal and plant species (IPCC 2007). Thus, to the extent that increased wind power development could appreciably reduce the severity of global climate change by reducing net greenhouse gas (GHG) emissions, the cumulative impact of wind power worldwide would provide a net benefit for biodiversity conservation. Nonetheless, the project-specific impacts of certain wind power facilities—wind farms, transmission lines, and access roads—can also be negative from a biodiversity standpoint. As detailed below, the adverse biodiversity-related impacts of wind power facilities mainly involve birds, bats, and natural habitats. Box 3.1 presents some of the reasons that wind power planners, developers, and operators are well advised to be concerned about biodiversity.

Box 3.1: Reasons to be Concerned about Biodiversity

In environmental terms and relative to other power generation options, wind power is unambiguously positive in terms of: (i) near-zero emissions of GHG and other air pollutants, (ii) negligible water consumption, and (iii) full renewability. Therefore, why should wind power planners or developers be concerned about the potential negative effects of wind turbines or associated infrastructure on birds, bats, or natural habitats? Some of the reasons include:

Aesthetic and Ethical Considerations. Aesthetic and ethical considerations underpin the concerns of many wind power stakeholders for generating and transporting electricity in ways that do not appreciably reduce the wild populations of particular bird and bat species, thereby helping to maintaining healthy, viable natural ecosystems.

Scientific and Economic Reasons. The birds, bats, and natural habitats that may be affected by wind power facilities have significant scientific and sometimes also economic value. Some insectivorous bats are of major (though not fully recognized) economic importance in consuming insects that are crop pests or nuisance species such as mosquitoes. In addition, some of the birds that are prone to collisions with wind turbines—such as eagles, storks, and other migratory species—are highly charismatic and of special interest for wildlife viewing; areas where these species are concentrated may be important from an eco-tourism standpoint. Finally, some of the windy sites that appear attractive for wind power development are also of considerable biological interest because of concentrations of migratory birds, other wildlife, or rare plants such as on mountain ridge-tops in the tropics.

Legal and Regulatory Requirements. Governments around the world have expressed their commitment to the conservation of birds and bats, among other wildlife, and protected natural areas through national and local laws, as well as international agreements such as the UN Convention on Biodiversity, Natura 2000 in the European Union, and several regional conventions on protecting migratory birds and other wildlife. Presently, wind project developers and operators face a variety of environmental legal and regulatory requirements in different countries, which sometimes involve potential legal liability for the incidental killing of protected bird or bat species by wind turbines (for example, the Migratory Bird Treaty Act and the Endangered Species Act in the United States). As the potential adverse impacts of wind power development on birds, bats, and natural habitats become better understood over time, the environmental legal and regulatory frameworks applicable to wind power might become more stringent in many countries.

Reputational Considerations. Although it has generally received less public attention in recent years than climate change, human-induced biodiversity loss is nonetheless widely recognized as a significant, ongoing global environmental problem with irreversible consequences such as species extinctions. Notwithstanding its other environmental advantages, wind power might not be perceived as truly "clean and green" if it is not also perceived as biodiversity-friendly—at least to the extent that any significant problems are effectively avoided and any adverse impacts are routinely and adequately mitigated. In this regard, the wind industry today enjoys generally favorable public opinion and supportive government policies in many countries, based in considerable measure on its highly positive reputation vis-à-vis the environment. Maintaining this positive reputation over time will require good management of all significant environmental and social issues, including the biodiversity-related ones.

Impacts of Wind Farms

IMPACTS ON BIRDS

Collisions with Wind Power Equipment. Bird mortality studies at wind farms (Orloff and Flannery 1992, Winkelman 1992, Marti and Barrios 1995, Thelander and Rugge 2000b, Leukona 2001, Erickson et al. 2001) indicate that most bird collisions are with the turbine rotor blades. These collisions most often kill the birds outright although they sometimes cause crippling. They do not damage the rotor blades or other turbine equipment. In addition, some birds are killed by colliding with the wind turbine towers themselves, including near ground level or with meteorological towers and their guy wires.

Birds' Awareness of Wind Turbines. Direct observations and radar studies of bird flight behavior in the vicinity of wind turbines show that on clear days and nights many birds actively avoid them, mostly by changing flight direction and flying around or away from the turbines (Koop 1997, Mossop 1997, Spaans et al. 1998, Guillemette et al. 1998, Janss 2000). Other studies describe birds flying unharmed through active turbine fields (Still et al. 1995, Spaans et al. 1998, van den Bergh et al. 2002, Smallwood and Thelander 2004; also, ongoing monitoring at the Mexico La Venta II wind farm). These data strongly indicate that the birds are generally aware of the turbines, and to some extent, recognize them as a potential hazard if approached too closely.

Reasons for Collisions. Nonetheless, significant numbers of birds do collide with wind turbines, apparently for two main reasons. First, while modern, large wind turbine blades appear to be moving rather slowly (15-20 rotations per minute), the blade tip speed is actually very fast: Typically about 7-10 times the mean local wind speed, up to about 270 kilometers per hour (kph), comparable to a jet plane right at takeoff. Therefore, the birds apparently do not perceive the danger until it is too late, and are struck by surprise (McIsaac 2001, Hodos et al. 2001, Hodos 2003), in a manner comparable to road kills from high-speed vehicles. Second, those birds that are actively looking for food around a wind farm—rather than just passing through—tend to be at greater risk because their attention is diverted away from the hazard posed by the spinning rotors. This behavior seems to be particularly important in raptors (birds of prey such as eagles, hawks, falcons, and owls) when they are focused on catching their prey and evidently inattentive to the turbines (Orloff and Flannery 1992).

Weather and Visibility Effects. Bird collision risks at wind farms tend to be greater at night and during inclement weather (Langston and Pullan 2002; Arnett, Inkley et al. 2007). Birds that migrate at night may simply not see the turning blades or even the towers. Inclement weather such as rain, fog, and strong headwinds forces many species of both diurnal and nocturnal migrants to fly at lower altitudes than normal, thus within the rotor-swept area (RSA) of operating turbines (Winkleman 1992, 1995). Collision risk is also elevated during very strong tail- or cross-winds when smaller birds in particular are buffeted about, lose much of their aerial maneuverability, and thus collide with the towers or rotor blades. This risk was noted during the Study Team's October 2008 visit to the Mexico La Venta II wind farm, when large numbers of migrating Scissor-tailed Flycatchers *Tyrannus forficatus* were observed, showing apparent difficulty in maneuvering around turbine towers during winds that exceeded 25 meters per second (m/sec). In addition, heavily wing-loaded soaring birds, such as Eurasian Griffon vultures *Gyps fulvus*, have low maneuverability and depend on lift to avoid wind turbines. Their collision

Photo: Carl G. Thelander

The Altamont Pass Wind Resource Area in northern California, USA, has become known for particularly high mortality of Golden Eagles and other raptors.

risk therefore increases in weather conditions where uplift winds, which are created by updrafts and thermals, are weak (De Lucas et al. 2007). Another factor in bird collisions with wind turbines is visibility. At night, collisions may be accentuated if birds are attracted to lights mounted on the wind turbine towers (Erickson et al. 2002).

Bird Species Most Significantly Affected. While all types of flying birds can collide with wind turbine blades or towers, some categories of birds are considered especially vulnerable because they: (i) tend to come in contact with wind turbines disproportionately often and/or (ii) have relatively low natural mortality rates, so that the additional mortality from wind turbines is more significant. For land-based wind farms, the bird groups of special concern include raptors, migratory species, and birds with aerial flight displays.

■   **Raptors.** Many species of raptors are highly vulnerable to collisions because they spend much of their time flying within the RSA height of the wind turbines. For example, at the Altamont Pass Wind Resource Area (APWRA) in California, USA (see Box 3.2), many of the approximately 70 Golden Eagles *Aquila chrysaetos* killed there by wind turbines each year were soaring and gliding within the RSA height while hunting rabbits and other small mammals that are abundant in the area (Smallwood and Thelander 2004). High raptor mortality rates—both in terms of absolute numbers and as a proportion of total bird mortality—have also been found at Smøla, Norway (Box 3.3), Tarifa in southern Spain (SEO/BirdLife 1995, Marti and Barrios 1995, Janss 2000), and Navarra in northern Spain. In Navarra, 227 dead Eurasian Griffons were found among 13 wind farms in 2000–2002 (Lekuona and Ursua 2007). This represents an unsustainably high mortality rate for a long-lived, slowly reproducing species with a total breeding population of about 2,000 pairs in Navarra and 20,000 pairs in all of Europe (EEA 2009). One Latin American species that seems to be highly collision-prone with wind turbines is the White-tailed Hawk *Buteo albicaudatus*, which spends a significant amount of time flying at RSA height while searching for small mammals and other prey. Monitoring at the World Bank-supported Mexico La Venta II project (Appendix A) suggests that the cumulative mortality from wind turbine collisions within the overall wind resource area exceeds local reproduction and that the local population is either in decline or is being maintained through influxes of birds from adjacent areas.

On the other hand, many other adequately monitored wind farms report very low raptor mortality, both in absolute and proportional terms, in European countries such as Germany (Durr 2003, Hotker et al. 2006), Belgium (Everaert et al. 2002), the Netherlands (Dirksen et al. 1998), the United Kingdom (Meek et al. 1993, Percival 2003), and Spain (Leukoma 2001). In the United States outside California (where raptor mortality is high, particularly at Altamont) and Texas (where wind farms are usually not monitored and/or the data are not disclosed), raptors have been reported to comprise only about 3 percent of turbine-related bird kills (Erickson et al. 2002, Kerlinger 2001).

- **Migratory Birds.** Migratory birds can be highly vulnerable to wind turbine collisions, particularly at night, during inclement weather, and when large flocks move through a wind farm. Wind farms located in areas with high concentrations of migratory birds tend to show correspondingly high collision rates. For example, high mortality in raptors and other large soaring birds such as White Storks *Ciconia ciconia* has also been documented at Tarifa, Spain, where thousands of birds concentrate at the Strait of Gibraltar each fall as they migrate from their breeding grounds in Europe to their wintering grounds in Africa (Marti and Barrios 1995). At the La Venta II project in Mexico, monitoring has revealed mortality of migrating birds (particularly songbirds that migrate at night), which is rather high by wind farm standards: 20 or more migratory birds may be killed per MW (megawatt) per year, based on the post-construction monitoring data (Appendix A) when likely correction factors (Appendix D) are taken into account. Nonetheless, this is a very small proportion of the several million birds that have been counted passing directly over or through the wind farm on an annual basis. Interestingly, more than half of the bird mortality at La Venta II is of resident (non-migrating) species. This appears to be because: (i) the resident birds spend much more time in a "high-risk neighborhood" than the migrants, and (ii) raptors and other birds appear to be more attentive to the turbines as a collision hazard when they are passing through than when they are hunting for food within the wind farm itself.

- **Birds with Aerial Flight Displays.** While some species of open-country birds (such as larks, pipits, and sparrows) forage for food well below the RSA of modern large turbines, the breeding males perform high-flying aerial courtship displays that cause them to spend large amount of time within the RSA (Kerlinger and Dowdell 2003). Consequently, these species often comprise a high proportion of the recorded bird mortality at monitored wind farms. While many of these species are very common and widespread, some are of increasing conservation concern, such as the declining Bobolink *Dolichonyx oryzivorus,* a long-distance migrant between the grasslands of North and South America, or many highly localized species of African larks.

Box 3.2: Raptor Mortality at Altamont Pass, California

Site Description: The Altamont Pass Wind Resource Area (APWRA) is located in northern California approximately 90 km east of San Francisco. Approximately 5,400 wind turbines, with a rated total capacity of 580 MW, are distributed over about 20,000 ha of grassy rolling hills and valleys. The APWRA is an important breeding ground, wintering area, and coastal migration corridor for a diverse mix of resident as well as migratory bird species. In particular, large numbers of hawks and eagles use the prevailing winds and updrafts for soaring and gliding during daily movement, foraging, and migration.

Bird Mortality Problem: Multiple long-term studies of bird mortality at Altamont Pass show that well over 1,000 raptors (birds of prey) are killed every year in turbine collisions, including about 67 Golden Eagles *Aquila chrysaetos*, 188 Red-tailed Hawks *Buteo jamaicensis*, 348 American Kestrels *Falco sparvarius*, and 440 Burrowing Owls *Athene cunicularia*, along with over 2,700 non-raptor species (Smallwood and Thelander 2008). Even though these species are protected by both federal and state wildlife legislation, bird mortality issues involving Altamont Pass are mostly addressed by Alameda County.

Mitigation Efforts: Attempts to reduce these high raptor losses, including poisoning campaigns for the rodents and rabbits that attract such high raptor densities, have been largely unsuccessful. Comparison of mortality data between studies conducted from 1998-2003 with those conducted from 2005-2007 actually showed marked increases in the mortality of most raptor species (Smallwood and Thelander 2009). In contrast to these results, which reflect the predominance of older, smaller turbines (40-400 kW) installed mostly in the 1980s, one wind farm within the APWRA that had larger, modern turbines showed a marked reduction in raptor and other bird mortality (Smallwood and Thelander 2009). These results suggest that bird mortality could be substantially reduced by installing modern high-capacity turbines in the WRA that are more widely spaced, strategically sited, and with the RSA ideally extending no lower than 30 m above ground. However, with the exception of two small portions of the WRA, to date this action has not been undertaken due to cost considerations combined with the lack of sufficient incentives or regulatory requirements.

Some Lessons:

1. The history of the Altamont Pass Wind Resource Area shows that environmentally careless choices made in wind farm location and turbine selection have long-term consequences. Even in a state known for generally strict environmental regulations and despite high levels of public support for conservation, it has been legally and politically problematic to implement mitigation measures that would prove costly for the wind farm operators. Unlike most types of energy or industrial development, land-based wind power in the United States is still regulated mostly at the local (county) level, rather than at the state or federal levels.

2. Key environmental mitigation actions—including those based on adaptive management in response to monitoring data—are much more likely to be implemented if they are provided for as part of the environmental licensing process, in advance of wind farm construction and operation.

Estimated Aggregate Bird Mortality from Wind Turbines. It is difficult to estimate total bird mortality from wind turbines because data are often lacking or of poor quality. At wind farms throughout much of the world, there is no systematic monitoring of bird or bat mortality because none is required by regulatory authorities. For example, Texas has the most installed wind power capacity of any U.S. state, but most wind farms do no such monitoring since it is not required at the county, state, or federal levels of government. In addition, many landowners and wind farm operators deny physical access to scientists interested in doing research, even at the scientists' own expense. Fur-

Photo: Rafael Villegas-Patraca
Black-bellied Whistling Ducks Dendrocygna autumnalis *fly through the La Venta II wind farm located in Mexico's Isthmus of Tehuantepec, a world-class bird migration corridor.*

thermore, at many other wind farms, while bird or bat monitoring data are collected, it is considered proprietary and therefore not disclosed to scientists or the general public. At some wind farms, the data collected are of poor quality due to methodological problems; therefore, bird mortality cannot be accurately compared between different wind farms. The greatest methodological issue is determining the appropriate correction factors for scavenger removal, searcher efficiency, and area not searched to account for the difference between observed bird or bat mortality and real mortality; the latter can be many times larger than the former (see the section "Post-Construction Monitoring" and Appendix D). Despite these limitations, some recent bird mortality studies have been conducted, in the United States and elsewhere, according to standard protocols (Morrison 1996, 2002a; Morrison *et al.* 2001). When monitoring follows these protocols, it can provide reasonable and directly comparable estimates of the numbers of birds killed by different wind farms.

Bird Conservation Significance of Wind Turbine Collisions. Wind power is still a very small proportion of all human-caused bird mortality. Erickson et al. (2005) estimated the average annual number of birds killed at U.S. wind farms that use standardized survey protocols and report their data to be 2.11 birds per turbine and 3.04 birds per MW. Based on the assumption that 3 birds are killed per MW per year, this would amount to about 75,000 birds killed by wind turbines in the U.S. during 2008. This number is in fact very small when compared with the estimated annual mortality of birds from collisions with buildings (especially glass windows), vehicles, telecommunications towers, outdoor domestic cats, pesticides, or hunting. Nonetheless, bird mortality at wind farms still merits serious attention for the following three main reasons:

- **Species of Conservation Concern.** For certain bird species, mortality at wind farms appears to be a relatively high proportion of total human-caused mortality. Many of these same species, especially raptors and certain other large, long-lived birds, have naturally low reproductive rates, so that any significant additional mortality could cause the population to decline. As an example, Smola Island, Norway had a dense, relatively stable population of White-tailed Eagles (19 breeding pairs) until a 68-turbine wind farm was built, after which the population rapidly collapsed to just one pair (Box 3.3). White-tailed Eagles and many other raptors are at little or no risk from buildings, vehicles, telecommunications towers, cats, or (depending on location) even pesticides or deliberate human persecution. Much the same could be said today of Golden Eagles in California (Box 3.2) or Eurasian Griffons in Spain. To date, no bird species has become globally endangered by wind power development. However, it is not inconceivable that the indiscriminate placement of new wind farms in areas heavily used by already threatened species—such as California Condors *Gymnogyps californianus* in the United States or Cape Griffon vultures *Gyps coprotheres* in southern Africa—could impede the species' recovery or even hasten their extinction.
- **Cumulative Impacts for Migratory Species**. If more wind farms continue to be built along bird migration pathways rather than lower-risk windy areas, the synergistic effect of multiple turbine risk zones experienced by migrating birds—in addition to all the other human-caused risks—may result in significant cumulative mortality.
- **Scaling Up Bird Mortality.** In the absence of effective mitigation measures (discussed below), the expected large-scale expansion of wind power capacity in many countries (by 1-2 orders of magnitude) is likely to lead to a corresponding increase in aggregate bird mortality.

Box 3.3: White-tailed Eagle Population Collapse at Smola, Norway

In 1992, Norway formed the state-owned Statkraft Group, now one of the largest producers of wind power in Europe. Statkraft's first wind farm was planned for Smola, a group of islands about 10 km off the northwest coast of Norway. This area had been designated an "Important Bird Area" (IBA) by the International Council for Bird Preservation in 1989 (Heath and Evans 2000), primarily because it had one of the highest densities of breeding White-tailed Eagles *Haliaeetus albicilla* in the world. This eagle is classified by BirdLife International (2000) as Globally Near-threatened, with an estimated world population of just over 10,000 birds. Norwegian and international conservation organizations had made repeated warnings to Norwegian environmental authorities that wind turbines at Smola would pose a significant threat to the eagles; they also filed a formal complaint with the Bern Convention (Norwegian Ornithological Society 2001). Nonetheless, the Government of Norway approved construction of the Smola wind farm, which began operation in 2002 and now consists of 68 (20 x 2 MW and 48 x 2.3 MW) turbines. Between 2005 and June 2010, 38 dead White-tailed Eagles have been found around the Smola turbines (Cole 2011); the number of breeding pairs on the island has dropped from 19 to only about one (BBC News 2006). The Smola experience demonstrates the importance of careful site selection for new wind projects, including an analysis of alternatives that takes a precautionary approach when assessing biodiversity implications.

The Greater Sage-Grouse Centrocercus urophasianus *(displaying male, left; typical habitat, right) of the western United States is among the open-country bird species that instinctively stay away from tall structures, including wind turbines and power poles.*

Displacement from Otherwise Suitable Habitat. For some bird species, the main risk posed by wind power development is not collision with the turbines, but displacement from their habitat by wind farms. Some bird species of open, naturally treeless habitats—including natural grasslands and shrub-steppe, perhaps also beaches and inter-tidal mudflats (Winkelman 1995; Kingsley and Whittam 2007)—are known to instinctively stay away from trees or any tall manmade structures, presumably based on an instinctive fear of hawks or other predators that use elevated perches. In North America, various species of prairie grouse are particularly prone to abandon otherwise suitable open-country habitat if any tall structures—trees, houses, grain silos, power poles, or wind turbines—are present (Manes et al. 2002). In Europe, Asia, and Africa, other shy birds of naturally open habitats, such as bustards, might similarly abandon extensive areas of otherwise suitable habitat around wind turbines despite the presence of otherwise suitable habitat.

IMPACTS ON BATS

Bat Kills at Wind Turbines. Recent studies have reported large numbers of bats being killed at wind farms in many parts of North America (Johnson 2005; Kunz, Arnett, Cooper et al. 2007; Arnett et al. 2008) and Europe (Ahlen 2002, Dürr and Bach 2004, Brinkmann et al. 2006); project monitoring has also discovered significant bat mortality at the Mexico La Venta II wind farm. Bat kills at wind turbines were first discovered in Australia (Hall and Richards 1972); small numbers of bats were first recorded in the

United States at wind power projects in California during bird monitoring (Orloff and Flannery 1992, Thelander and Rugge 2000b). More recently, an estimated 1,400-4,000 bats were recorded as killed during 2003 at the Mountaineer Wind Energy Center in West Virginia (Kerns and Kerlinger 2004); high bat mortality at that site has continued since then (Arnett et al. 2008). The frequency and number of bat kills at wind turbines are much greater than for any other type of human-built structure; unlike birds, bats strike telecommunications towers, buildings, or power lines only very infrequently (Avery and Clement 1972, Crawford and Baker 1981, Mumford and Whitaker 1982).

Bat Mortality from Collisions and Barotrauma. Bats that fly too close to wind turbines are killed by either direct impact or from major air pressure changes around the spinning rotors. While bats clearly are killed by direct collision with turbine blades (Johnson et al. 2003, Kerns et al. 2005, Baerwald et al. 2008), up to 50 percent of the dead bats around wind turbines are found with no visible sign of injury (Kerns et al. 2005, Baerwald et al. 2008). The cause for this non-collision mortality is believed to be a type of decompression known as barotrauma, resulting from rapid air pressure reduction near moving turbine blades (Kunz, Arnett, Cooper et al. 2007; Baerwald et al. 2008). Barotrauma kills bats near wind turbines by causing severe tissue damage to their lungs, which are large and pliable, thereby overly expanding when exposed to a sudden drop in pressure (Baerwald et al. 2008). By contrast, barotrauma does not affect birds because they have compact, rigid lungs that do not excessively expand.

Bat Attraction to Wind Turbines. Many species of bats appear to be significantly attracted to wind turbines for reasons that are still poorly understood (Horn et al. 2008a). Box 3.4 summarizes the more plausible scientific hypotheses that have been advanced to date. By contrast, birds are not normally attracted to wind turbines, and simply collide with them by accident.

Photo: Ed Arnett, Bat Conservation International
The Eastern Red Bat Lasiurus borealis is typical of the migratory, tree-roosting bat species that are frequent casualties at some wind farms in North America.

Box 3.4: Hypotheses for Bat Attraction to Wind Turbines

Various scientific hypotheses have been proposed as to why bats are seemingly attracted to and/or fail to detect wind turbines (Arnett 2005; Kunz, Arnett, Cooper et al. 2007; Arnett et al. 2008; Cryan 2008; Cryan and Brown 2007; Horn et al. 2008a; Lacki et al. 2007). The more plausible hypotheses include the following:

Auditory Attraction: Bats may be attracted to the audible "swishing" sound produced by wind turbines. Museum collectors seeking bat specimens have used long poles that were swung back and forth to attract bats and then knock them to the ground for collection; it is not known if these bats were attracted to the audible "swishing" sound, the movement of the pole, or both factors (M. D. Tuttle, Bat Conservation International, *pers. comm.*).

Electromagnetic Field Disorientation: Wind turbines produce complex electromagnetic fields, which may cause bats in the general vicinity to become disoriented and continue flying close to the turbines.

Insect Attraction: As flying insects may be attracted to wind turbines, perhaps due to their prominence in the landscape, white color, lighting sources, or heat emitted from the nacelles, bats would be attracted to concentrations of prey.

Heat Attraction: Bats may be attracted to the heat produced by the nacelles of wind turbines because they are seeking warm roosting sites.

Roost Attraction: Wind turbines may attract bats because they are perceived as potential roosting sites.

Lek Mating: Migratory tree bats may be attracted to wind turbines because they are the highest structures in the landscape along migratory routes, possibly thereby serving as rendezvous points for mating.

Linear Corridor: Wind farms constructed along forested ridge-tops create clearings with linear landscapes that may be attractive to bats.

Forest Edge Effect: The clearings around wind turbines and access roads located within forested areas create forest edges. At forest edges, insect activity might well be higher, along with the ability of bats to capture the insects in flight. Resident bats as well as migrants making stopovers may be similarly attracted to these areas to feed, thus increasing their exposure to turbines and thus mortality from collision or barotrauma.

Thermal Inversion: Thermal inversions create dense fog in cool valleys, thus concentrating both bats and their insect prey on ridge-tops.

Bat Species Most Significantly Affected. In North America, migratory bat species have been found dead at wind farms much more frequently than the resident (non-migratory) species, even in areas where the resident species are more common throughout the summer (Johnson *et al.* 2003, Arnett *et al.* 2008). Eleven of the 45 species of bats that occur in North America north of Mexico have been found dead at wind farms (Johnson 2005), but most studies report that the mortality is heavily skewed towards migratory, tree-roosting species such as Hoary Bat *Lasiurus cinereus,* Eastern Red Bat *Lasiurus borealis*, and Silver-haired Bat *Lasionycteris noctivagans* (Johnson 2005; Kunz, Arnett, Cooper et al. 2007; Arnett et al. 2008). While these three species are not listed as threatened or endangered under the U.S. Endangered Species Act, they are classified as of Special Management Concern at the provincial level in Canada. Although the globally endangered Indiana Bat *Myotis sodalis* has not yet been found dead at wind farms, potential new wind farms within this species' remaining strongholds could possibly threaten it (Woody 2009).

In Europe, 19 of the 38 species of bats found within the European Union have been reported killed by wind turbines (L. Bach, unpublished data). Although migratory species are among the most numerous casualties, resident bats are also killed in substantial numbers, particularly in forested areas (Dürr and Bach 2004). Turbine-related bat mortality has been found in every European country in which bat monitoring has been done, except for Poland where no dead bats were found during monitoring at two sites (L. Bach, unpublished data). The highest numbers of bat fatalities have been found in Germany and France, which is almost certainly due to the more extensive monitoring carried out in those countries (Behr et al. 2007).

In Latin America, 19 bat species were represented among the 123 individual bats found dead under wind turbines in 2007-2008 at the La Venta II project in southern Mexico; in 2009, 20 different bat species were involved (INECOL 2009). Thirteen of these species are insectivores, while two feed mainly on nectar, and two on fruit. The most commonly killed species, Davy's Naked-backed Bat *Pteronotus davyi*, is thought to be resident in the area, although some other frequently killed species at La Venta II are considered to be migratory. Interestingly, despite the enormous concentrations of migratory birds that pass over or through the La Venta II wind farm (over 1 million per year), monitoring data from INECOL show that a larger number of bats are being killed there than birds. At the World Bank-supported Uruguay Wind Farm project, systematic post-construction bat monitoring is expected take place for three years beginning in 2010. Preliminary monitoring trials found turbine-related mortality of one bat species, the widespread Brazilian Free-tailed Bat *Tadarida brasiliensis*. At the Colombia Jepirachi Wind project, a site with sparse desert vegetation and lacking fresh water, bats have not been monitored; however, no bats (alive or dead) were evident or reported during the World Bank's environmental and social supervision visit in August 2007. As of this writing, it is not known if any other bat monitoring is being carried out at wind farms in Latin America, or anywhere else outside of North America and Europe.

Higher Wind Turbine Mortality for Bats than for Birds. Based on the monitoring carried out to date, bats appear to be killed by wind turbines at significantly higher rates than birds, except at those sites where bats are naturally scarce or absent. Reported kills of bats almost always exceed, often by a factor of 2-4, those for birds at nearly all wind farms studied to date (Erickson et al. 2001, 2002; Barclay et al. 2007; Arnett et al. 2008). Estimates of bat mortality from 21 studies located at 19 different wind farms from 5 different regions in the United States and Canada ranged from 0.9-53.3 bats/MW/year, with a mean of 12.8 bats/MW/year (95 percent confidence interval 5.7-19.9, Arnett et al. 2008). This compares with the estimate by Erickson et al. (2005) of 3.04 birds/MW/year at U.S. wind farms. The significantly higher mortality for bats than for birds may be explained, at least in part, because: (i) bats appear attracted to wind turbines (rather than encountering them only by chance) and (ii) barotrauma from decompression means that, unlike birds, bats can be killed just by closely approaching an operating wind turbine, without even touching it.

Bat Conservation Significance of Wind Turbine Mortality. Because bats are long lived and have exceptionally low reproductive rates, population growth is relatively slow, and their ability to recover from population declines is limited (Barclay and Harder 2003). Despite the small size of most bat species, these animals tend to be more like raptors and seabirds in their potential vulnerability to the added mortality from wind turbines than most faster-reproducing small bird species. In this regard, wind farms can become a local "population sink" for certain bat species, a situation in which mortality exceeds reproduction and the local population is maintained through influxes from adjacent areas. This might possibly be the case at the Mexico La Venta II Project (Appendix A). The projected rapid scaling up of wind power development could lead to significant bat population declines over more extensive areas. For example, if the United States. were to achieve the goal of 20 percent wind power generation by 2030 (DOE 2008), corresponding to an increase from 25,000 MW in 2008 to 300,000 MW in 2030, over 4 million bats would be killed annually, assuming a continuation of the average estimated mortality of 12.8 bats/MW/year (Arnett et al. 2008). Although bats can be counted at large cave roosts, there are no good census methods—nor population estimates—available for the tree-roosting bat species that are most frequently killed at North American wind farms (Carter et al. 2003). Nonetheless, it is possible that the populations of some bat species are too small to withstand the projected future annual mortality from wind turbines (Racey and Entwistle 2003; Kunz, Arnett, Cooper et al. 2007; Winhold et al. 2008). In the absence of adequate mitigation measures, large-scale wind power expansion in much of the world could quite possibly threaten the survival of certain bat species with distributions that overlap heavily with wind farm installations.

IMPACTS ON NATURAL HABITATS

When located in areas of natural vegetation, wind farms can sometimes harm biodiversity through the clearing and fragmentation of natural habitats. The footprint of cleared natural vegetation from wind turbine platforms and interconnecting roads tends to be around 1-2 ha/MW). For example, at the Uruguay Wind Farm, the 27 ha of natural grassland vegetation that were mostly cleared during turbine installation are being used for five turbines of 2 MW each, plus a small house and substation. From a biodiversity conservation standpoint, this extent of clearing may be significant in specific cases, such as ridge-top forests. Mountain ridge-tops—particularly in the tropics—can have unique animal and plant life, often due in part to their wind-swept micro-climate. Long rows of turbines with interconnecting roads along such ridge-tops can disproportionately affect scarce, highly localized species. For example, proposed wind farms, including access roads, along windy, forested ridge-tops in Panama such as within the Santa Fe National Park threaten to remove a significant proportion of the unique ridge-top vegetation that harbors endemic species of plants and animals, including the endangered Glow-throated Hummingbird *Selasphorus ardens* and highly localized Yellow-green Finch *Pselliophorus luteoviridis*. Another harmful impact of wind farms can occur if they reduce surface-level wind speeds enough to alter downwind sand dune formations, thereby potentially affecting the survival of endemic, dune-adapted plant and animal species; this type of risk has been identified by scientists for several sand dune areas in Wyoming, USA (Molvar 2008).

Impacts of Complementary Infrastructure

POWER TRANSMISSION LINES

Wind Projects Require Transmission Lines. Most wind power projects require the installation of power transmission lines—sometimes of considerable length—to connect with the national or regional power grid. As with other power generation technologies, wind turbines are economically useful only when they are physically connected to their electric power market. While this chapter focuses on the biodiversity impacts associated with wind turbine facilities (wind farms), it also briefly addresses the impacts of power transmission lines since they are an essential component or pre-condition of any wind power project. Even though there are distinct environmental advantages to burying power lines, overhead transmission lines still predominate in most of the world due to their technical simplicity and lower construction costs. In most countries, overhead, rather than buried, transmission lines are used to connect wind farms and other types of power plants with major power demand centers. However, power lines are typically buried within the wind farm itself, which is good practice with respect to worker safety, visual impacts, and avoiding bird mortality. The main potential adverse biodiversity-related impacts of power transmission lines include:

- **Bird Collisions.** Many species of birds collide with overhead transmission lines, particularly with the top (grounding) wire that is least visible. Particularly vulnerable to collisions are relatively fast-flying, heavy-bodied birds with limited maneuverability during level flight. Such birds include pelicans, cormorants, herons, storks, ibises, flamingoes, waterfowl (ducks, geese, and swans), cranes, bustards, and coots. Power line collisions are usually most problematic over wetlands since these habitats tend to have high bird concentrations in general, but particularly for most of the above-mentioned species groups with high collision vulnerability. However, the best-documented case of a bird species that is actually believed to be threatened with extinction because of power line collisions is the Ludwig's Bustard *Neotis ludwigii*, endemic to the semi-arid Karoo plains of South Africa and Namibia, an area increasingly criss-crossed with power transmission lines (Jenkins and Smallie 2009). Overhead power transmission lines are still much more widespread than wind farms; they are believed to kill many more birds overall than turbines (Erickson et al. 2005). Unlike birds, bats do not normally collide with power lines or other structures besides wind turbines.

- **Bird Electrocutions.** Bird electrocutions most typically involve medium or large birds that perch on medium-voltage lines or power poles and complete a circuit by simultaneously touching two live wires, or a live wire and grounding element. Large raptors (including eagles, hawks, and vultures) are particularly vulnerable to such electrocutions, which can be the largest source of human-caused mortality in some areas; one study estimates that 52 percent of the Harris' Hawk *Parabuteo unicinctus* mortality near Tucson, Arizona (USA) was due to electrocutions (Alison 2000). Power line electrocutions have also caused alarmingly high levels of mortality among threatened Spanish Imperial Eagles *Aquila adalberti* and Bonelli's Eagles *Hieraaetus fasciatus* in Europe (CMS 2002a), as well as Cape Griffons and African White-backed Vultures *Gyps africanus* in South Africa (Van Rooyen 1998).

- **Other Bird Interactions with Power Lines.** Overhead power transmission lines (including bird-friendly designs that involve little or no electrocution risk) are often used by raptors as perches for resting and hunting. In certain instances, small mammals or birds of conservation concern may suffer increased predation from raptors that are attracted to perch on the power lines or poles. For this reason, the strategic environmental assessment (SEA) of wind power in Wyoming (Molvar 2008) recommends that transmission lines attached to wind power projects avoid passing through or near Black-tailed Prairie Dog colonies, which also harbor the endangered Black-footed Ferret. Furthermore, various bird species tend to build nests on the convenient platforms provided by certain designs of power poles. Depending on the power pole configuration, nesting at these locations can be problematic because of electrocution risks for the birds (such as raptors in Mongolia, discussed by Harness and Gombobaatar 2008), as well as electrical short-circuits that can interrupt the power supply. In many other cases, power pole configurations are bird friendly and do not pose electrocution risks to nesting birds; which is the case with the 16 km transmission line for the Uruguay Wind Farm Project.

- **Forest Fragmentation.** Under power line rights-of way (ROW), tall trees are not allowed to prevent physical interference with the lines and possible electrical shorts. When a transmission line—from a wind farm or any other power generation facility—crosses forested areas, the result can be significant deforestation that incrementally adds to the project's carbon footprint. The forest fragmentation resulting from the power line ROW can also be problematic from a biodiversity standpoint: This situation occurs in the case of power transmission lines in the Amazon Basin of South America where several species of monkeys, among other wildlife, spend their entire lives high in trees and never cross open ground. The cumulative result of multiple power lines—coupled with roads, pipelines, railways, and other linear clearings—can thus be genetically isolated populations of vulnerable species, potentially reducing their long-term survival prospects.

3.1.2.2 Access Roads

For some wind power projects, the most significant adverse environmental impacts can result from road construction and improvement rather than from the wind turbines or transmission lines. Wind power projects frequently require: (i) new access roads to reach the wind farm site; (ii) major widening of existing roads to facilitate the passage of very large trucks, hauling rotor blades, nacelles, and turbine tower sections; and (iii) new roads to connect all of the turbines on a wind farm. Certain wind projects also build or improve access roads to benefit local communities even if these roads are not needed for wind farm construction or operation. The biodiversity-related and other environmental impacts resulting from road construction or major improvement can include: (i) direct impacts resulting from the road works themselves, as well as (ii) induced (indirect) impacts resulting from new or increased human activity that is made feasible by the improved road access. Among others, the direct impacts include: (i) loss of natural habitats from clearing for the road ROW, (ii) fragmentation of forests or other natural habitats, (iii) altered drainage patterns of waterways, (iv) soil erosion and landslides, (v) pollution and sedimentation of aquatic ecosystems, (vi) airborne dust, (vii) blockage of fish and wildlife movements, and (viii) wildlife road kills. The induced impacts of new or improved access roads tend to involve socioeconomic drivers and are often more severe and/or difficult to mitigate, particularly in developing countries. They can include: (i) new human settlement of previously remote areas; (ii) deforestation or other clearing or degradation of natural habitats; (iii) logging and other wood cutting, often at unsustainable levels; (iv) informal mining development and the associated environmental impacts; and (v) illegal or unsustainable hunting or harvest of vulnerable species (Ledec and Posas 2003, Ledec 2005).

Biodiversity Impacts of Offshore Wind Development

Birds. Certain biodiversity-related environmental issues are either less problematic or absent at offshore wind power sites; however, some new issues arise. With respect to birds, some offshore areas are heavily traversed by migratory landbirds and shorebirds. For instance, migrating raptors favor relatively narrow water crossings between two points of land; one such example in the U.S. is between Cape May, New Jersey and the nearby Delaware shoreline. Higher-flying species of seabirds and waterfowl are at risk of turbine collisions (Dierschke et al. 2006, Garthe and Huppop 2004), while various species can be displaced (scared away) by turbines from key offshore feeding areas (Guillemette et al. 1999, Tulp et al. 1999, Larsson 1994, Dirksen et al. 1998). For some water birds, a wind turbine array may become a barrier to movement between offshore feeding areas and onshore nesting sites (Langston and Pullan 2002, Exo et al. 2003, Hotker et al. 2006, Kingsley and Whittam 2007, Percival 2001). As with landbased wind power, careful site selection for offshore wind farms is a key consideration in minimizing these adverse impacts.

Bats. Migratory bats fly across stretches of ocean, apparently much like migrating landbirds (Cryan and Brown 2007). Interestingly, some 10 bat species (resident as well as migrating species) have been studied feeding on insects well off the coasts of Sweden and Denmark, at heights ranging from the water surface up to the around the top of the turbine RSA (Ahlen et al. 2007). Presumably, similar behavior also occurs off the coasts of many other countries. Nonetheless, offshore wind farms likely pose a much smaller threat to bat populations than do land-based facilities because onshore bat populations and species diversity are much higher overall.

Marine Life. Marine mammals tend to be driven away by the loud noise and vibration from pile driving and other works during wind farm construction. For example, noise levels of 70 dB(A) (decibels corrected or "A-weighted" for sensitivity of the human ear) were noted 46 km away during pile driving for construction of the Alpha Ventus offshore wind farm in Germany, causing all marine mammals to abandon the area (R. Gebhardt *pers. comm.* 2010). However, marine mammals generally return and resume using the same offshore habitat during project operation (EEA 2009). The hard substrate around wind turbine platforms and the submerged tower bases themselves create a "reef effect" that attracts fish and other marine life. This reef effect is a generally positive impact, although certain fish species might be disadvantaged by changes in underground habitat and increased competition from species that favor hard substrates (EEA 2009). An important, environmentally positive impact of offshore wind farms is that the exclusion of fishing boats creates *de facto* marine protected areas, which is beneficial to fish and other marine life but can also generate local opposition from fishing interests.

Biodiversity-Friendly Practices for Wind Power Projects

Wind Project Planning

SITE SELECTION OF WIND POWER INFRASTRUCTURE

The Importance of Site Selection. Wind power infrastructure includes wind turbines, meteorological towers, access roads, and transmission lines. Careful site selection of wind power infrastructure is often the most important measure in managing the biodiversity and other environmental impacts of wind projects. To a very considerable degree, site selection can predetermine the biodiversity-related impacts, as well as extent to which follow-up mitigation measures might be effective.

Higher-Risk Sites. Experience with wind power projects to date suggests that, from a biodiversity conservation standpoint, the more problematic sites for wind power infrastructure tend to have one or more of the following characteristics listed below. However, since much still remains to be learned about the factors that predispose birds, bats, or other biodiversity to risk at wind farms, this list should be regarded as only a "first approximation." Depending on specific circumstances, wind project planners who encounter one or more of these characteristics are advised to either search for an alternative project site, or carry out more detailed due diligence to ensure that the biodiversity-related and other environmental and social risks are manageable.

- **Protected Areas.** Around the world, many types of national parks, wildlife refuges, forest reserves, and other protected areas have been established to conserve in perpetuity their biodiversity, outstanding natural beauty, and/or physical cultural heritage. In many existing and proposed protected areas, the construction of a wind farm or other large-scale energy development facility would be considered incompatible with the main reasons for having set aside the protected area in the first place. Nonetheless, certain categories of protected areas, within certain countries, do allow for the installation of wind farms — as well as hydroelectric, geothermal, oil and gas, or other energy development. Often efforts to locate wind farms within protected areas are highly controversial with environmental nongovernmental organizations (NGOs) and other stakeholders; a case in point involves recently proposed wind farms within Panama's Santa Fe National Park and San Lorenzo Protection Forest, which were vigorously opposed by local NGOs.

- **Critical Natural Habitats.** In general terms, critical natural habitats (CNHs) comprise sites of known high biodiversity conservation value, whether or not they are located within existing or officially proposed protected areas.[1] CNHs include areas that are considered important for the survival of endangered, vulnerable, rare, and migratory species. Various country-level lists and maps of CNHs have been prepared by government agencies or conservation NGOs. Particularly useful, in terms of identifying sites that would best be avoided for wind power development, are the Important Bird Areas (IBAs) that have been designated for many countries by the NGO BirdLife International and its partner organizations.

- **Wildlife Migration Corridors.** For migratory birds, many of the key migration "bottlenecks," staging areas, and concentration points are already well known;

in some countries, these sites may be visited by large numbers of birdwatchers. For migratory bats, there is comparatively little knowledge of areas where bats tend to concentrate during migration, although shorelines, mountain ridges, and/or mountain passes are likely possibilities. Migration corridors are also important for the survival of many populations of large mammals that need to walk long distances to reach seasonal food supplies; examples include Elk *Cervus elaphus* in western North America, as well as many species of grazing animals in African savannas. For these typically shy creatures, wind turbine towers *per se* do not seem to be the problem; rather, it is the near-constant presence of wind farm employees and their vehicles during project operation and especially during construction (Molvar 2008). The necessary due diligence for wind farm site selection is most effective when it is done in consultation with government agencies and conservation NGOs, who are likely to have information about important wildlife migration corridors—whether or not they lie within existing or proposed protected areas.

Photo: Bats and Wind Energy Cooperative (BWEC)
Wind turbines on forested mountain ridges can pose a relatively high risk to bats, as well as specialized ridge-top vegetation.

■   **Wooded Areas.** Forests and woodlands frequently hold higher bat populations than more open habitats, particularly for the tree-roosting bat species that are killed at wind turbines in disproportionately high numbers (Arnett et al. 2008). As noted above, forest clearing and fragmentation to install wind turbines and associated roads (roughly 1-2 ha/MW) can adversely affect a variety of animal and plant species, particularly those that are naturally restricted to mountain ridge-tops. Moreover, clearing wooded land for wind power production or transmission produces incremental greenhouse gas emissions when the felled biomass is burned or decomposes. For these reasons, wind farms installed in forested areas tend to be less environmentally desirable than most wind farms located in open areas; the same can be said of power transmission lines. From a biodiversity standpoint, clearing native forest is more problematic than clearing forest plantations of non-native species.

- **Wetlands.** Wetlands comprise a wide variety of natural or modified, permanent or seasonal, systems of fresh, brackish, salty, or alkaline waters; they include, among others, marshes, wooded swamps, inter-tidal mudflats, lakes (natural or impounded), lagoons, ponds, rice paddies and other flooded fields, streams, rivers, bays, and estuaries. Wetlands are considered high-risk sites for locating wind turbines because they tend to attract concentrated populations of birds such as waterbirds, raptors, and many other species and bats that come to feed on the concentrations of insects. For example, Hötker et al. (2004) show that nearly half of the German wind farms studied had mortality of less than one bird/turbine/year, but that a few wind farms had mortality exceeding 50 birds/turbine/year; the high-mortality wind farms were located either in wetlands or on mountain ridge-tops. Similarly, Jain et al. (2007) found a significant negative relationship between the number of dead bats and the distance from wetlands, thus showing that wind farms located closer to wetlands tend to kill more bats. Wetlands also tend to be high-risk sites for overhead power transmission lines because these ecosystems tend to harbor sizable bird concentrations, especially of aquatic bird species that are highly prone to power line collisions. Furthermore, aside from bird or bat collisions, wetlands can be environmentally problematic sites for installing wind farms because turbine access roads, if built on top of dikes, can adversely affect drainage patterns and wetland hydrology, sometimes over large areas (Ledec and Posas 2003); roads on dikes can also block the migrations of fish and other aquatic fauna or plant life of conservation interest (Ledec 2005).

- **Shorelines.** The shorelines of oceans, seas, or large lakes (including cliff-tops) tend to be windy, but they also tend to have concentrated bird and bat populations, including: (i) seabirds, waterfowl, and shorebirds that forage along the coast; (ii) migrating birds and bats that tend to follow coastlines; and (iii) migrating birds or bats that concentrate near the shore, just before or after a long water-crossing flight. For these reasons, shorelines inherently tend to be higher-risk sites for locating wind turbines. For those wind farms that are nonetheless established near the coast, adequate setback from the shoreline can greatly reduce bird mortality, often with little or no reduction in favorable wind speeds. At the World Bank–supported Colombia Jepirachi Wind Power project (Appendix B), the minimum setback from the shoreline was 200 m, as a precautionary measure to minimize seabird mortality; this measure almost certainly has reduced wind farm mortality of Brown Pelicans *Pelecanus occidentalis,* and probably of other seabirds and shorebirds as well.

- **Small Islands.** Small islands can be high-risk sites for installing wind power facilities, particularly if they either contain nesting seabird colonies or lie along major bird or bat migration pathways. Small islands located within bird migration corridors tend to serve as a "magnet" for large numbers of birds that are flying over a large expanse of water, see the island, and stop to rest and feed.

■ **Native Grasslands or Shrub-Steppe.** These naturally open ecosystems merit careful biological review during wind project site selection because they might be important habitats for shy species of birds (section "Impacts on Birds") or large mammals that could be displaced (frightened away) from important areas of habitat.

■ **Near Caves.** Caves and similar underground structures (such as abandoned tunnels and mine shafts) sometimes harbor large numbers of breeding or roosting bats. Certain wind farms in Texas, USA have been installed relatively close to large breeding colonies of Brazilian Free-tailed Bats, although no monitoring of bat mortality has been carried out there because the wind farm operators have not allowed it, and Texas law does not require it. However, significant bat mortality seems likely at these particular wind farms, based on data obtained from just across the state line in Oklahoma (M. D. Tuttle, Bat Conservation International, *pers. comm.*).

Lower-Risk Sites. Conversely, as a first approximation, lower-risk sites for wind power development tend to have low bird and bat numbers year-round, and do not harbor animal or plant species of conservation or special management concern. Such lower-risk sites often fall into one of the three categories below. Nonetheless, specific screening of these sites as part of the EA process is still very important. The assessment of biodiversity risks from wind power facilities, especially for flying birds and bats, still faces a steep learning curve, so there is always the potential for surprise findings, even at previously assumed low-risk sites.

■ **Most Cultivated Lands.** Extensive areas under cultivation are less likely to harbor bat or bird concentrations, or species of conservation interest. In terms of minimum tradeoffs with biodiversity, such lands can be ideal sites for wind farm development. However, there are some exceptions to this general rule, including: (i) some grain fields (especially near wetlands) that may seasonally attract large migratory bird flocks (especially irrigated rice, which is a type of human-created wetland), and (ii) irrigated fields in desert areas, which due to the local abundance of fresh water, serve as oases that can attract large numbers of birds and other fauna, such as near the Salton Sea in southern California, USA.

■ **Non-native Pastures.** Pasture lands with predominantly non-native grasses or other forage plants tend not to harbor important populations of the grassland bird species of greatest conservation concern, which could be harmed by collisions or habitat displacement. However, some, though not most, non-native pasture lands are located within important migratory bird corridors or in areas of high raptor densities, as is the case at Altamont Pass, California.

■ **Deserts.** Open, barren desert lands tend to have relatively low bird and bat densities, and are thus potentially well-suited for wind power development. Important exceptions to this rule include those desert areas with freshwater oases or near coastlines, where bird numbers can be quite high.

WIND FARM LAYOUT AND MICRO-SITING

Wind Farm Configuration. Wind farms are sometimes arranged as a single row of turbines, or as multiple rows of varying length. The studies carried out to date do not suggest a very strong relationship between wind farm configuration *per se* and bat or bird mortality. For example, at wind farms in the eastern U.S. (Kerns et al. 2005, Fiedler 2004, Fiedler et al. 2007) and Alberta, Canada (Brown and Hamilton 2002, 2006 a, b), bat mortality was about evenly distributed between all turbines, regardless of their position within the wind farm. However, having a relatively wide space between turbines is likely to be useful for migratory birds, by making it easier to avoid turbine collisions when passing through a wind farm at RSA height (Smallwood and Thelander 2004). Accordingly, a "wind wall" design, in which a wind farm's turbines are rather closely clustered within a mountain pass, is likely to be riskier to migratory birds and ideally would be avoided. In landscapes that are a mosaic of forested and agricultural lands, it is usually preferable to avoid locating turbines within the forested areas, both to conserve biodiversity and to further reduce GHG emissions.

Micro-Siting of Wind Turbines. In terms of minimizing bird and bat collisions or other harm to biodiversity, wind farm configuration *per se* seems to be less important than the careful location of individual turbines to minimize proximity to specific, sensitive sites within the wind farm area. Such sites can include raptor nests, specific areas of high bird or bat use, or the habitats of species of conservation concern. These sites should be identified and demarcated prior to construction, as part of the EA process. For example, at Foote Creek Rim, Wyoming, USA, pre-construction surveys showed that about 85 percent of raptors flying at RSA height were within 50 meters of the canyon rim edge. As a result, the project sponsor agreed to move back the nearest row of turbines from the rim edge by 50 m, although conservationists had recommended 100 m (Molvar 2008). Post-construction monitoring at Foote Creek Rim confirms relatively low raptor mortality—about 0.04 raptors/MW/year—versus about 2.2 raptors/MW/year (55 times higher) at Altamont Pass (Young et al. 2003). At the same wind farm, turbines were located to avoid damaging a colony of burrowing Black-tailed Prairie Dogs *Cynomys ludovicianus*, which also contain the endangered Black-footed Ferret *Mustela nigripes*.

For proposed wind farm development in the Argentina's Patagonia region, minimum setback distances have been suggested from those landscape features where birds tend to be most numerous, including protected areas, bird nesting colonies, inland or coastal wetlands, landfills, and slaughterhouses that attract scavenging birds. Special care should be taken to configure turbine arrays so as to avoid creating a barrier between bird nesting and feeding areas; at Zeebrugge, Belgium, high mortality was recorded among breeding terns that had to cross a line of wind turbines situated between their nesting colony and regular feeding areas (Everaert and Stienen 2006). Within a general wind farm area, specific turbines can be located so as to minimize the number and length of new roads, when desired to minimize the loss and fragmentation of natural habitats. The placement of each turbine should also take into account the efficiency of power generation based on prevailing winds and the need to minimize generation losses in the wake of other turbines.

Within a particular wind resource area, the biodiversity and other environmental risks of wind power development can be minimized through a process of phased development. Under this approach, one wind farm (comprising between about 5 and 50 turbines) is installed, and then systematically monitored for environmental impacts, particularly bird and bat mortality. If the monitoring shows these impacts to be acceptably low, then scaling up with additional wind farms is warranted in nearby areas with similar characteristics. This phased approach is often financially feasible, and is sometimes chosen for non-environmental reasons. For example, the Uruguay Wind Farm project involves only five turbines (10 MW), but another five turbines have been proposed as an adjacent, follow-up future development. In the meantime, planned monitoring will verify whether, as expected, the existing five-turbine wind farm can be operated without significant harm to birds or bats. In general, a phased approach—involving a relatively small pilot phase with intensive monitoring, followed by an expanded development phase—helps to prevent large-scale, essentially irreversible mistakes in wind power site selection. In retrospect, using such a phased development approach at high-risk sites such as Altamont Pass could have prevented the substantial mortality of eagles and other raptors that still occurs today.

WIND POWER EQUIPMENT SELECTION

The choice of wind power equipment—turbines, met towers, transmission lines, and even lighting fixtures—can substantially affect biodiversity outcomes, particularly with respect to bird and bat collisions.

Wind Turbines

Turbine Design. The dominant wind turbine design in current use is the horizontal axis, propeller-type design, generally with three rotor blades. Several types of vertical axis wind turbines have also been produced for smaller-scale, off-grid uses. However, the vertical-axis turbines tend to be more expensive to manufacture because they require a greater volume of materials to build, relative to the useful wind area captured. Vertical-axis turbine designs are often promoted as "wildlife friendly," and their appearance strongly suggests that bird and bat collisions would be less likely (Box 3.5). However, Anderson et al. (2004) found that vertical axis turbines of the FloWind type used at Tehachapi Pass, California had similar bird mortality rates to the conventional, horizontal axis turbines. Field studies have not yet been done to compare bat mortality in relation to turbine design.

Turbine Size. For commercial-scale wind power production, the current trend is for wind turbines to become progressively larger—up to 2 or 3 MW per turbine for most land-based wind projects, and around 5 MW for offshore turbines. It is often difficult to install turbines larger than about 3 MW on land because rotor blades longer than 45-50 m can be difficult to transport by road. There are some notable exceptions, such as the Enercon E126 turbines at Estinnes, Belgium that are being upgraded to 7 MW (R. Gebhardt, *pers. comm.* 2010). Larger turbines tend to be higher off the ground, have a larger RSA, and be spaced further apart within wind farm rows, as well as between rows. Although longer rotor blades turn more slowly in terms of revolutions per minute (RPM)—15-20 RPM vs. 60+ RPM for smaller turbines—their rotor tip velocity is about as fast as for the smaller blades (around 270 kph, as discussed in the section "Impacts on Birds").

Box 3.5: Alternative Turbine Designs

Wind turbines in use today include two basic design types: Horizontal-axis wind turbines (HAWTs; Figure 1) and vertical-axis wind turbines (VAWT; Figures 2 and 3). HAWTs have the main rotor shaft and electrical generator at the top of a tower and must be perpendicular to the prevailing wind. VAWTs have the main rotor shaft arranged vertically, do not have to be pointed into the wind, and the generator and gearbox can be placed near the ground so it does not need a tower for support. Wind speeds near the ground tend to be relatively low. Turbines used in wind farms for commercial production of electric power are usually three-bladed horizontal-axis turbines. This design is highly efficient as it allows access to stronger winds in areas where winds increase with height above the ground. In addition, it also usually has variable pitch blades that give the turbine blades the optimum angle of attack to collect the maximum wind. The Darrieus "eggbeater" turbine (Figure 2) and the Savonius scoop turbine (Figure 3) were tried and tested in early generation wind farms, but were generally found to be less cost-effective than the HAWT design.

Figure 1. Conventional Horizontal-axis turbine (HAWT)

Figure 2. Darrieus "eggbeater" turbine

Figure 3. Savenius scoop turbine

Overall, larger wind turbines appear to be less risky for birds and bats than the smaller ones, at least on a per-MW basis. Most species of birds, including raptors and bats, fly less than about 30 m (100 feet) above the ground when foraging for food, so they are more likely to fly beneath, rather than through, the RSA of the largest land-based turbines (Smallwood and Thelander 2004, Hoover and Morrison 2005). Moreover, since larger turbines need to be spaced further apart in wind farm rows, birds may be more likely to find their way safely between individual turbines. Some experts in bird and bat collisions with wind turbines consider that the probability of collision is largely a function of the absolute number of turbines, more so than turbine size (C. Thelander, R. Villegas-Patraca, and E. Arnett, *pers. comm.*); by this line of reasoning, a smaller number of larger turbines will kill fewer birds or bats (all else equal) than a larger number of smaller turbines. Although empirical data that test these hypotheses are scarce, intensive monitoring of bird mortality at Altamont Pass, California indicates that the most important available measure to reduce raptor mortality while operating would be to repower with the largest feasible turbines since most raptors as well as other birds would then be flying below the RSA most of the time (Altamont Pass Avian Monitoring Team 2008).

Smaller turbines might be preferable in the special case of birds and bats that migrate at night, typically around 100-600 m above ground level. For these species, the upper RSA of the larger turbines might encroach into their flight paths, especially near mountain passes, during inclement weather, and while ascending or descending (Barclay et al. 2007). However, when wind turbines are located away from major migration corridors (as they ideally should be), a smaller number of large turbines is generally preferable to a larger number of smaller ones in terms of reducing bird (and perhaps also bat) collisions.

On a per-MW basis, larger wind turbines also have other environmental advantages over smaller ones. Less land area needs to be cleared of its vegetation since fewer turbine platforms are needed. Given that large turbines can be visible for a long distance in any case, a smaller number of turbines appear to have less of a visual impact on a per-MW basis. For the Uruguay Wind Farm project, the power utility UTE chose to install five 2-MW turbines, instead of twelve 800 kW turbines, in part because local residents indicated that the visual impacts would be less severe.

Turbine Color. Presently, almost all large wind turbines (towers, nacelles, and rotor blades) worldwide are painted white or off-white, sometimes with red stripes on the rotor blades and corporate logos on the nacelles. Since bats navigate primarily by echolocation[2] (Griffin 1958), color is not considered to be a relevant factor in efforts to reduce turbine-related mortality. For birds flying at night or in very dim light, white appears to be a good color choice because darker turbines would be less visible under these conditions. However, during daylight hours, other colors or patterns would generally be more visible to birds, particularly when the turbines are viewed with a background of open sky.

To date, little research has been done on whether other colors or patterns could significantly reduce bird collisions with wind turbines. With the exception of ultra-violet (visible to birds but not people), using other colors to make turbines more visible to birds would also make them more visible to people—a generally undesirable outcome, except in remote or special-use areas such as military bases where visibility issues are not a major concern. For the Gulf of El-Zayt wind farm project in Egypt (financed by KfW [Germany], the European Investment Bank, and the European Commission), the turbines to be installed will each have two white and one black rotor blades, in an effort to make them more visible to large numbers of migrating White Storks *Ciconia ciconia* (R. Gebhardt, *pers. comm.* 2010). Even when rotor blades are painted with a more highly visible color or pattern, data are still lacking as to whether bird collisions can be significantly reduced. As noted in the section "Impacts on Birds," birds appear to be struck because of the rotor blades' surprisingly fast speed, rather than because their color is not the most conspicuous.

The strongest case for considering alternative turbine colors might be to paint the lower base of the turbine tower a darker color, while leaving the rest of the tower white. Although most bird mortality at wind turbines involves collisions with the rotor blades, a surprising number of birds are killed by flying into the turbine tower itself, often very close to the ground. For example, at La Venta II (Mexico), a notable proportion of the bird carcasses found at wind turbines are of non-migratory birds that tend to stay close to the ground and never fly as high as the RSA. These birds include the Northern Bob-white *Colinus virginianus*, Stripe-headed Sparrow *Aimophila ruficauda*, and the highly lo-calized endemic Cinnamon-tailed Sparrow *Aimophila sumichrasti*. On multiple occasions, bird monitoring staff at La Venta II have observed Northern Bobwhites flushing from the ground and flying directly into the turbine towers—very probably because from below, the broad, white turbine tower bases look rather like open sky. For birds that collide with the low stratum of wind turbine towers, a feasible solution might be to paint the bottom 3-5 meters of each tower a dark color so that it more clearly looks like an obstacle, rather than open sky. Such a small, dark base to a very tall (50-60 m) tower would mostly be visible within the wind farm itself rather than from further away where aesthetics-relat-ed visibility issues would more frequently arise. Some wind farms in northern Germany follow this approach, with dark green bases to their otherwise white turbine towers.

Wind Farm Lighting. Wind farm lighting includes: (i) aircraft warning lights that are installed on tops of some turbine nacelles and (ii) night lights around wind farm buildings, parking lots, or other facilities. Aircraft warning lights on top of tall telecom-munications towers are known to be highly problematic for night-flying migratory birds (though not bats) during cloudy weather when the birds cannot navigate via the stars and are attracted to the lights, thereby colliding with the towers, guy wires, or each other.

Recent U.S. research indicates that white strobe lights with brief pulses are more bird friendly than solid or slowly pulsating red or white lights, since they seem to be less of an attractant to migrating birds on cloudy nights (Longcore et al. 2005, Erickson et al. 2005). The windy sites most favorable for wind farm development often, though not always, have clear skies at night when night-flying birds are not attracted to artificial lights.

To minimize the risks that night-flying migrant birds will be attracted to wind tur-bines during overcast weather, aircraft warning lights would ideally be: (i) white strobe lights and (ii) placed atop a limited number of turbines to help show the wind farm outline, rather than on every one, consistent with national and local regulatory require-ments. In many low-risk cases far away from airports, such lights might not be needed at all; for example, U.S. law requires aircraft warning lights only on structures 200 feet or higher. If aviation regulations provide flexibility regarding the type of aircraft warning lights to be used, it would also be important to consult with local residents regarding their aesthetic preferences—while also explaining the bird-friendly advantages of white strobe lights. At the Mexico La Venta II project, only a small number of the 98 turbines have aircraft warning lights, which appears to be adequate for showing any pilots that might be in the area that there are multiple structures around, without potentially at-tracting numerous migrating birds to the turbines on overcast nights.

Night lighting close to ground level can also be managed to minimize bird and bat mortality. Night lights around worker compounds, parking lots, and other wind farm facilities can attract to the wind farm migratory birds during cloudy weather, as well as bats and nocturnal birds (owls and nighjars) that seek out the flying insects that con-centrate around night lights. In this regard, wind farm night lighting can be managed to minimize risks to birds and bats in ways that do not compromise worker safety and operational security. These approaches include the use of: (i) sensors and switches to keep lights off when not needed and (ii) lighting fixtures that are hooded and directed downward to minimize the skyward and horizontal illumination that could attract night-flying birds and bats to the vicinity of wind turbines.

Perchable Structures. Birds (particularly raptors) are known to perch on various wind turbine structures, especially on the lattice-work towers that were widely used in the past, but have largely been replaced in recent years with smooth, tubular towers. The modern, tubular towers are widely regarded as more bird friendly than the lattice towers since raptors frequently perch on the latter, in close proximity to hazardous spinning rotor blades. However, even tubular towers have some structures on which birds can perch. Some of these structures are considered indispensable, such as the anemometer and safety railings atop the nacelle. Other external structures are not needed if good alternatives exist; a case in point is external ladders, which are not needed when steps or elevators are inside the turbine tower. To minimize risks to birds, turbine towers would have only those perchable structures that are clearly needed for operational efficiency or worker safety. Another important aspect of wind turbine maintenance is to ensure that turbine structures do not have any open holes in which birds would try to roost or nest (see the section "Wind Farm Maintenance Practices" and Appendixes 1 and 3). Unlike birds, bats are not known to use perchable structures on wind turbines.

Ultrasound Speakers. Many species of bats heavily use echolocation for orientation, prey capture, communication, and avoiding obstacles. Because echolocating bats depend upon sensitive ultrasonic hearing, broadcasting high-intensity sounds at a frequency range to which bats are sensitive could mask echo perception, or simply create an uncomfortable or disorienting airspace that bats might seek to avoid (Spanjer 2006, Szewczak and Arnett 2006). Research on ultrasonic sound emission as a possible deterrent to bats is underway in the United States. (E. Arnett, Bat Conservation International, unpublished data). Currently, there is still no effective alerting or deterring mechanism that has been proven to effectively reduce bat (or bird) mortality at wind turbines. For deterrents to be effective, they must operate at ranges that are large enough to encompass an area greater than the turbine RSA. Thus, if an effective sound-based deterrent were to be developed in the near future, it would most likely be successful on relatively small turbines.

Meteorological Towers

Bird Collisions with Guy Wires. Meteorological—also called met towers or masts—continually measure wind speed and direction at potential and operational wind farm sites. To minimize costs, most met towers are relatively thin structures, held in place by numerous guy wires that are often visible—to birds and people—only at close range. Unlike the wind turbines themselves, met towers are not considered a problem for bats, which are not attracted to them and successfully manage to avoid them. However, met towers with guy wires often kill birds because: (i) the guy wires can be nearly invisible to flying birds, particularly in fog, dim light, or at night; and (ii) birds that are buffeted by high winds often lack the maneuverability to avoid the guy wires at the very last moment. An example of relatively high mortality is from the Foote Creek Rim, Wyoming wind farm where the number of birds found dead at five guyed met towers exceeded the number killed at all 69 turbines by about a factor of 4 (Young et al. 2003).

Mitigation Options. One option for reducing bird mortality is to install met towers without using guy wires, as has been done at some wind farms, such as Nine Canyon in Washington State, USA. However, because un-guyed met towers need to be sturdier than guyed towers to remain standing in very strong winds, they are considerably more expensive to procure and install. Moreover, the installation requires greater use of heavy machinery and sometimes the construction of additional access roads. Nevertheless, un-guyed met towers merit consideration for those wind power projects located along bird migration pathways or in other areas with large concentrations of birds, particularly when species of conservation concern are involved.

If guyed met towers are used, the most important mitigation option may be sim-ply to ensure that the minimum number of towers is left standing, particularly once the wind farm is operational. Even though every modern wind turbine has its own an-emometer to measure wind speed and direction, met towers tend to measure ambient wind conditions more reliably, which facilitates efficient wind farm operation. However, wind project developers and operators can help to minimize bird mortality by ensuring that only those guyed met towers that are important for proper wind farm functioning remain standing. Typically, this means having one, occasionally two, towers per wind farm.

Transmission Lines

Key Considerations in Equipment Selection. As noted earlier, the most important tool for mitigating the biodiversity and other environmental impacts of overhead power transmission lines is usually careful alignment, to avoid crossing the most sensitive ar-eas. With respect to equipment selection, the most important considerations from a bio-diversity standpoint are: (i) bird flight diverters and (ii) bird-friendly power pole and line configurations, described briefly below. More detailed information on these options is available from the U.S. Avian Power Line Interaction Committee (APLIC), www.aplic.org.

■ **Bird Flight Diverters.** Bird flight diverters (BFDs) are devices that can be at-tached to overhead transmission line grounding and conducting wires in order to make them more visible to birds. One well-known flight diverter design is a large, typically orange, ball that is placed on transmission line grounding wires, usually near airports and at canyon and river crossings to alert airplane pilots. Various other BFD designs exist in different shapes and colors. Some flap in the breeze, while others are completely stationary. BFDs are commercially avail-able under brand names that include Firefly and BirdMark. In general, BFDs are inexpensive to install when the power line is first under construction, but much more expensive to retrofit. If installed properly, BFDs do not affect elec-tricity transmission performance (APLIC 1994). BFDs are effective in reducing many bird collisions, as De La Zerda and Rosselli (2003) demonstrate in their detailed case study from a wetland in northern Colombia. Accordingly, the rou-tine use of BFDs is highly recommended when constructing or replacing over-head transmission lines in higher-risk areas, particularly wetlands. However, some bird species, such as the threatened Ludwig's Bustard in southern Africa, seem less responsive to BFDs and continue to collide with transmission lines at high rates (Jenkins and Smallie 2009). In these cases, the environmentally preferred solution might be to re-route transmission lines around the species' most important habitat and/or follow existing transmission line or other linear infrastructure corridors in order to avoid affecting new areas of habitat.

■ **Bird-Friendly Power Pole Configurations. Cer**tain configurations of medium-voltage power poles, power lines, and associated equipment (insulators, transformers, and so forth) are considered bird friendly because they pose little or no electrocution risk; these are described in detailed and well illustrated by APLIC (2006). In general, bird-friendly configurations are those in which: (i) electrified wires are at least 2-3 m apart (greater than the wingspan of the largest local raptor species) if at the same height and (ii) insulators, transformers, and similar devices either point down (not up!) from the cross-pole where a bird might perch, or are otherwise outside the normal stretching reach of a perched bird.

ENVIRONMENTAL PLANNING TOOLS

A variety of environmental planning tools are available to optimize the location and design of wind power projects in terms of biodiversity and other environmental, social, and economic criteria. These tools include strategic environmental assessments as well as project-specific environmental assessments with environmental management plans. In addition, well-written bidding documents and contracts can help ensure environmentally more favorable wind power equipment and construction practices.

Strategic Environmental Assessments

Strategic environmental assessments (SEAs) are a key tool for wind power planning, providing information that would not normally be covered in project-specific environmental impact assessments (EIAs). SEAs are known by different names, including sectoral and regional environmental assessments, and vary greatly in their scope and level of complexity. In terms of wind power planning, the important functions that SEAs perform include: (i) macro-level site selection (within a country or region) of the optimum wind resource area (WRA) or sub-region for developing new wind farms and power transmission corridors; (ii) assessing the cumulative environmental impacts of multiple wind farms within a WRA; (iii) developing common good practice standards for use by all wind power developers within a WRA (such as for post-construction monitoring, operational curtailment, or social benefits-sharing); (iv) developing mechanisms for information exchange and shared responsibilities between different wind farm operators; (v) analyzing alternative power generation options (in addition to wind) within a particular planning area; and (vi) providing a platform for involving different stakeholders, including the most vulnerable, in the decision-making process regarding wind development. Some specialized forms of SEAs focus on particular issues such as the cumulative impacts of multiple wind farms within a WRA on migratory birds or other wildlife. This type of study is also called a cumulative impact assessment. Certain SEAs address social opportunities, constraints, and risks at the regional level, in effect making them the integrated tool known as a strategic environmental and social assessment (SESA). Governments interested in scaling up wind power could help optimize environmental and social—as well as economic—outcomes by promoting SEAs in areas with high wind potential well in advance of issuing wind development concessions and permits.

Overlay Maps. As discussed above, proper site selection of wind power facilities is normally the most important measure available for minimizing adverse biodiversity and other environmental impacts. Since potential wind power resources remain largely untapped in most countries, they typically have multiple options regarding where to locate new wind farms, in terms of the connection to a national or regional electricity grid. For macro-level site selection of new wind power development, SEAs provide overlay maps that show where the zones of high wind power potential based mainly on wind speeds and proximity to the power transmission grid are located, in relation to the areas of major environmental and social sensitivity. Environmentally or socially sensitive areas are likely to include, among other features: (i) protected areas and other sites of concern from a biodiversity standpoint, (ii) areas important for tourism where visual impacts would be of concern, (iii) areas with uncertain or disputed land ownership, (iv) areas with indigenous or other traditional rural populations where greater-than-usual efforts might be needed to design culturally appropriate benefits-sharing measures and obtain broad community acceptance, (v) radar and telecommunications facilities where turbines could cause interference, and (vi) areas close to airports.

Zoning Maps. In addition to overlay maps that show areas of overlap and potential conflict, some SEAs also produce zoning maps for prospective wind power development, recommending such zones as: (i) "Red" Exclusion Zones, where wind farms or transmission lines would be prohibited; (ii) "Yellow" Precautions Zones, where wind farm development would need to follow special precautions based on the specific natural or cultural resource(s) of concern; and (iii) "Green" Promotion Zones, where wind farm development could be actively promoted, subject to the standard environmental and social due diligence, or perhaps prescreened for expedited approval. Aside from minimizing environmental and social conflicts, a coordinated planning approach with zoning maps can facilitate wind power development by identifying areas where the electric grid will require expansion or reinforcement.

SEA for Wyoming, USA. As an example, an SEA for the U.S. state of Wyoming was recently completed by a consortium of conservation NGOs (Molvar, 2008). The report, entitled *Wind Power in Wyoming: Doing it Smart from the Start,* identifies: (i) 10 different categories of existing and proposed protected areas; (ii) raptor nesting concentration areas and Bald Eagle *Haliaeetus leucocephalus* roost sites; (iii) wooded areas of likely importance to bats; (iv) key birthing areas, migration corridors, and winter feeding grounds for large mammals including Pronghorn *Antilocapra americana*, Elk, Mule Deer *Odocoileus hemonius*, Bighorn Sheep *Ovis canadensis*, Mountain Goat *Oreamnos americanus*, and Moose *Alces alces*; (v) Black-tailed Prairie Dog colonies; (vi) Black-footed Ferret recovery areas; (vii) Greater Sage Grouse *Centrocercus urophasianus* and Sharp-tailed Grouse *Tympanuchus phasianellus* leks (courtship sites); (viii) Mountain Plover *Charadrius montanu* (nesting concentration areas); (ix) sand dune ecosystems; (x) National Historic Landmarks, National Historic Trails, and other important historic sites; and (xi) cities and towns. For each of these sensitive site categories, the report applies transparent criteria to classify them as recommended Red or Yellow zones, with buffer areas of appropriate width. Based on these recommendations, of the 25 percent of Wyoming with wind conditions suitable for commercial development (Classes 4-7), 41 percent is zoned as Red (no-go), 48 percent is Yellow (special precautions), and 11 percent is Green (wind power promotion area).

SEA for Tehuantepec, Mexico. To help optimize the siting and minimize the adverse cumulative environmental impacts of multiple future wind farms and transmission lines within the south Tehuantepec wind resource area (WRA), an SEA is planned as a component of the GEF-funded Large-Scale Renewable Energy (La Venta III) project.

For this WRA, the SEA is expected to produce: (i) a zoning map indicating which sites would be less sensitive (and which would be more problematic) with respect to wind farm and power line development, based on biodiversity and other environmental criteria; (ii) recommendations on how to minimize adverse cumulative impacts in the siting of wind farms, transmission lines, and access roads; and (iii) recommended standard bird and bat monitoring techniques for future wind farms with adjustments for higher and lower-risk sites. In addition, a separate study has been proposed to prepare a national-level "environmental overlay" of areas of known high sensitivity (for example, protected areas, critical natural habitats, and areas with high bird or bat concentrationsand so forth) on the same scale as the planned National Wind Resource Map. In this manner, the mapped areas of high wind resource potential could readily be compared in terms of their likely degree of environmental sensitivity to help inform future governmental and private sector decisions about where best to locate future wind power investments within Mexico.

Environmental Impact Assessments

Environmental impact assessment (EIA) reports for wind power projects represent an essential tool for identifying and properly managing site-specific biodiversity and other environmental impacts. While an SEA can identify relatively broad zones of environmental suitability, the project-specific EIA is the appropriate instrument for a more detailed, micro-level screening of specific wind farm sites, and even of specific turbine placements within a wind farm. Moreover, the EIA report typically documents and explains the criteria and methods that were used for wind project site selection.

As the biodiversity and other environmental impacts of wind power development become better known, more governments are likely to require EIAs for wind projects. While some governments currently require EIA reports for wind power projects, others do not. In those cases where EIA reports are required for other kinds of large-scale energy development but not for wind, it might be due to the still widespread perception that wind power projects are "clean energy" and thus lack any significant adverse impacts. In some countries, EIA studies are required only above a certain wind project size; for example, EIA studies in Uruguay are required only for projects exceeding 10 MW. EIA studies were carried out for all three of the Latin American case study projects for this report (in Mexico, Colombia, and Uruguay) since it was a World Bank requirement; however, only in the case of Mexico was the EIA also required by the government.

Pre-Construction Biodiversity Studies. As part of the project-level EIA process, it is often important to carry out specialized, site-specific studies of particular bird or bat species, species groups, or natural habitat areas that may be of special concern in the context of a proposed new wind project. Surveys of resident or (in-season) migratory birds within the proposed wind project area can provide a variety of information useful for decision-making, such as: (i) the relative importance to birds (particularly, of species of conservation interest) of the proposed project site in relation to other potential wind power sites within the country; (ii) the extent to which birds might be adversely affected by the project, either through collisions (if they frequently fly within the RSA height) or habitat loss or degradation; and (iii) particular sites such as raptor nests, bird breeding colonies, roosting trees, or staging areas that should be specifically avoided when planning the layout of individual turbines, access roads, substations, worker camps, transmission lines, or other project facilities. In the case of bats, obtaining information on sites where bats might be expected to concentrate in flight—near caves or other known roosts, as well as likely feeding areas such as around wetlands and forest edges—can be helpful in wind farm site selection and perhaps also micro-siting of individual turbines. However, in comparison to birds, pre-project surveys for bats have not been proven effective at predicting where large numbers of migratory bats may potentially become wind turbine casualties (Kunz, Arnett, Erickson *et al.* 2007). Additional detailed guidance on specific methods and protocols for these types of pre-construction bird and bat surveys is available from Anderson et al. (1999); Strickland et al. (2009); NWCC (2007); Canadian Wildlife Service (2007); Kunz, Arnett, Cooper et al. (2007); CEC (2007, 2008); and Molvar (2008).

For pre-construction surveys of bats and nocturnal birds, specialized equipment and techniques are typically needed. These equipment and techniques include acoustic detection (such as with Anabat and other hand-held devices), night vision observations (including spotlights, infrared cameras, and night-vision goggles), thermal infrared imaging, Doppler weather surveillance radar (NEXRAD), tracking radar, and marine (ship) radar. Kunz, Arnett, Cooper et al. (2007) provide a comprehensive description of each technique, their overall utility, and general cost. At the Mexico La Venta II project, a marine radar unit (costing about US$20,000) has been mounted to the back of a pickup truck, and is being effectively used for both spotting approaching daytime bird flocks and for counting nocturnally migrating birds. Marine radar has also been successfully used at several proposed and operational wind farms in the United States (Harmata et al. 1999, Cooper and Day 2004, Mabee and Cooper 2004, Desholm et al. 2006, Mabee et al. 2006). Compared with NEXRAD and tracking radars, marine radar is relatively inexpensive, available off-the-shelf, highly portable, dependable and easy to operate. It also requires little modification or maintenance, has repair personnel readily available worldwide, and has high resolution in both the horizontal and vertical scanning modes.

Environmental Management Plans. EIA reports for wind power projects are most useful if they include some type of environmental management plan (EMP); sometimes other names are used) that specifies the actions to be taken during project construction and operation to prevent, minimize, mitigate, or compensate for any adverse environmental impacts, and to enhance any positive impacts. For each recommended mitigation, enhancement, monitoring, and/or training activity, the EMP would ideally specify: (i) an implementation schedule in relation to construction activities and turbine operation, (ii) the institutional responsibilities for carrying out the environmental management actions and supervising compliance, (iii) the corresponding one-time and recurrent costs, and (iv) the dedicated source(s) of funds to cover these environmental management costs.

Bidding Documents and Contracts. As with any type of infrastructure or energy development project, obtaining the desired biodiversity and other environmental outcomes does not end with a good EIA report. In this regard, the key standards for equipment selection, project construction, and wind farm operation (as specified in the EMP) need to be incorporated into all relevant bidding documents and contracts.

Equipment Procurement Specifications. In recent years, the rapid worldwide growth in wind farm development has led to a "sellers' market" for wind turbines and related equipment. If wind developers wanted to avoid long delays, often they could not afford to be too selective with respect to turbine size or other technical specifications. However, as the wind power market matures over time, it is likely that wind project developers will have more leeway in choosing equipment specifications. Accordingly, project procurement plans should take into account those equipment specifications that would be the most environmentally benign in terms of biodiversity and other criteria based on recommendations included in the EMP. As discussed above, the more biodiversity-friendly types of wind power equipment tend to involve: (i) turbines that can be programmed for different cut-in speeds at different hours of day or night; (ii) turbines that can be feathered on demand in real time; (iii) larger rather than smaller turbines; (iv) turbines with the minimum feasible perchable structures; (v) turbine towers with the bottom 3-5 meters painted a darker color; (vi) white strobe lights (rather than solid or slowly pulsating red or white lights) as aircraft warning lights on the tops of turbine nacelles or met towers; (vii) low-level lighting fixtures that are hooded, point downward, and have sensors and switches; (viii) transmission lines with bird flight diverters; and (ix) power line poles and insulators that are dimensioned and positioned to avoid bird electrocutions.

Wind Project Construction

Environmental Rules for Contractors. As discussed above, adverse biodiversity and other environmental impacts can be greatly reduced through the careful choice of project location (wind farm site, specific turbines within the wind farm, access road and transmission line alignments) and type of wind power equipment (turbines, met towers, and transmission lines). Adverse impacts can be further minimized through the use of good environmental practices during the construction of wind farms and their associated access roads and transmission lines. These good practices include: (i) minimizing any clearing of natural vegetation during turbine installation; (ii) locating worker camps, storage sheds, parking lots, and other construction-related facilities so as to avoid or minimize the removal of natural vegetation, opting instead to use previously cleared or degraded lands; (iii) installing sufficient drainage works under all access roads, to avoid flooding land and damaging streams; (iv) implementing adequate measures to control soil erosion and runoff; (v) ensuring proper disposal of solid and liquid wastes; (vi) refraining from washing of vehicles or changing of lubricants in waterways or wetlands; (vii) ensuring that locally obtained construction materials (such as gravel, sand, and wood) come from legal and environmentally sustainable sources; (viii) following chance finds procedures if archaeological, historical, or other relics are unearthed during project construction; (ix) restoring cleared areas where feasible to minimize the wind project's environmental footprint, although the area around each turbine should typically have minimal or short ground cover for the duration of post-construction monitoring of bird and bat mortality (sections "Post-Construction Monitoring" and "Wind Farm Maintenance Practices"; Appendix D); and (x) enforcing good behavior by construction workers, including prohibition of hunting, fishing, wildlife capture, bushmeat purchase, plant collection, unauthorized vegetation burning, speeding, firearms possession (except by security personnel), or inappropriate interactions with local people (Ledec and Posas 2003, Ledec 2005).

   Getting Results on the Ground. Bidding documents and contracts for turbine installation, road construction, and any other civil works should specify the key environmental management measures during construction so that all contractors and subcontractors are aware of them and can adjust their construction budgets and schedules accordingly. Furthermore, to help ensure that environmentally friendly plans on paper are actually implemented, strict field supervision of construction work needs to be carried out by independent, knowledgeable personnel: "You get what you *inspect*, not what you *expect*" (G. Keller, U.S. Forest Service, *pers. comm.*). In this regard, project sponsors and construction supervisors also need to ensure that financial or other penalties such as termination or black-listing are sufficiently strict to promote good compliance by contractors and construction workers and enforced in a transparent manner.

Wind Project Operation

Following wind farm site selection, the operation of wind projects usually provides the best opportunities for managing biodiversity-related impacts. During project operation, bird and bat monitoring are a key element of good environmental management. Also of particular significance for bird and bat conservation are some potentially cost-effective options for operational curtailment. Furthermore, the careful management of wind farm landscapes can significantly enhance biodiversity outcomes. All of these good operational practices are most likely to be implemented if they are explicitly described and budgeted within signed project agreements and contracts; in this regard, wind farm operators are understandably often reluctant to implement practices that have not been agreed upon up front.

POST-CONSTRUCTION MONITORING

The highest-priority post-construction monitoring of biodiversity impacts most often involves estimating bird and bat mortality when the wind farm is operational. However, post-construction monitoring of biodiversity impacts can involve a variety of additional activities. For example, post-construction monitoring at the Mexico La Venta II project also includes: (i) daytime observation of migrating bird flocks, their numbers by species, and the routes they follow; (ii) use of a mobile ship radar, mounted on the back of a pickup truck, to detect migratory bird flocks by day and night; (iii) systematic observations of how birds behave when flying through the wind farm, including turbine avoidance reactions; and (iv) monitoring of local bat and (non-migratory) bird populations, including their habitat use within the wind farm, bird nesting activity, and searches for caves and other bat roosting areas.

Reasons to Monitor Bird and Bat Mortality. Searching for the carcasses of project-killed birds and bats at wind farms can seem grim and unpleasant; it also does not present the most favorable public image of the wind power industry, which is widely perceived as being "clean" and "green." Nonetheless, this type of monitoring is so important from an environmental and scientific standpoint that it would ideally be standard practice for all wind turbine projects. For the three Latin American case study projects highlighted in this report, post-construction monitoring of bird and bat mortality is being carried out in a thorough manner at Mexico La Venta II; was done, but with limited effort and inconsistent record-keeping at Colombia Jepirachi; and, at this writing, is getting underway with robust plans at the Uruguay Wind Farm. Post-construction monitoring is a key element of the environmental management of wind power projects for the reasons noted below.

- **Knowing Whether a Problem Exists.** Irrespective of the extent of pre-construction environmental studies, the only way to know for sure if a wind project is causing significant bird or bat mortality is to look for carcasses adjacent to the turbines. Pre-construction assessments can overestimate, underestimate, or completely overlook the risks to particular groups of birds or bats. For instance, at La Venta II, where a significant pre-construction assessment of risks to migratory birds was carried out as part of the environmental assessment (EA) process for the project, many surprising findings have emerged as a result of post-construction monitoring. These include: (i) short-term, real-time turbine shutdowns to protect large migratory bird flocks are very feasible, but needed only on rare occasions (mainly during unusual weather conditions); (ii) more bats appear to be killed than birds; (iii) resident birds have higher turbine-related mortality overall than migratory species; (iv) an unexpectedly high number of very low-flying birds are colliding with the bases of turbine towers (not the rotor blades); and (v) American Kestrels *Falco sparverius* attempt (hazardously) to nest or roost within nacelle holes.

- **Predicting Impacts from Scaling Up.** Within a specific wind resource area, the documented bird or bat mortality at an operating wind farm is generally the best available predictor of mortality at a proposed new facility. Thus, the bird and (especially) bat post-construction monitoring that is planned for the 5-turbine, 10 MW Uruguay Wind Farm will be very useful in assessing the likely impacts from a proposed future doubling of the size of this facility. Similarly, the data made available by post-construction monitoring at the Mexico La Venta II project are very useful in anticipating the impacts of new wind farms within the south Isthmus of Tehuantepec, even though some portions of this WRA are more heavily used than others by migratory birds.
- **Adapative Management of Wind Farm Operation.** Data on bird or bat mortality are needed to allow for informed decision-making about whether and how wind farm operation should be modified to reduce such mortality, through some type of operational curtailment (higher cut-in speed or short-term shutdowns), equipment maintenance, or landscape management.
- **Advancing Scientific Knowledge.** As should by now be evident to the readers of this report, the science of how wind power affects wildlife still faces a steep learning curve. In this regard, proper collection, analysis, and dissemination of bird and bat mortality data from wind power facilities are essential to the timely advancement of scientific knowledge about the biodiversity impacts of this rapidly growing technology.

Key Aspects of Monitoring Bird and Bat Mortality. At its most basic level, post-construction monitoring for bird and bat mortality involves searching for, identifying, counting, and analyzing the bird and bat carcasses found close to wind turbines, met towers, and/or associated transmission lines. Because bird and bat mortality can vary from year to year, especially due to weather conditions, it is recommended that post-construction monitoring be carried out for at least the first two years of wind farm operation—longer if those two years are considered atypical in climatic terms, such as during the El Niño/La Niña weather phenomena. If, after this trial monitoring period, bird and bat mortality are found to be insignificantly low, then the monitoring can be discontinued, or perhaps reduced in intensity to occasional "spot checks." However, if significant bird and bat mortality is found, then the monitoring should continue so that operational curtailment or other mitigation measures can be implemented, tested, and refined. To cover most of the area in which turbine-killed birds or bats would fall, the search area should encompass a radius equivalent to the maximum RSA height around each turbine to be searched. Furthermore, to minimize any conflicts of interest (real or perceived), post-construction biodiversity monitoring needs to be contracted to an independent entity, rather than carried out directly by the wind farm operator or project sponsor. Further, detailed guidance on how to carry out bird and bat carcass searches is provided by Strickland et al. (2009); Kunz, Arnett, Cooper et al. (2007); Smallwood (2007); CEC and DFG (2007); PGC (2007); Anderson et al. (1999); and Morrison (2002a).

Real versus Observed Bird and Bat Mortality. In the course of post-construction bird and bat monitoring, it is essential to recognize that there is a very real—and sometimes very large—difference between real mortality and observed mortality at wind farms. Due to this difference, it is important to use the best available correction factors that take into account the area not searched, searcher efficiency, and scavenger removal of carcasses. As explained in Appendix D, the difference between real and observed mortality for large raptors and other very large birds is likely to be relatively small. However, for small birds and especially bats, the difference between real and observed mortality can be very large—perhaps a factor of 50 at a project such as La Venta II in Mexico.

Automatic Collision Detection. New automatic strike detection technology, currently under development, would significantly improve the accuracy of mortality estimates for birds and bats at wind turbines. One of the simpler and inexpensive designs involves the use of acoustic microphones with a laptop computer processor that can be installed, or retrofitted, onto turbine nacelles. Sometimes called a "thunk detector," this device records the sounds of any objects of a particular size and density (i.e. birds or bats) that strike the rotor blades. While this device might become available for commercial use in the very near future, turbine manufacturers, wind project sponsors, and electric utilities would then need to be persuaded to install it. In addition, more complex and expensive systems that use thermal imaging or surveillance cameras are also under development (W. Evans, *pers. comm.*).

Data Compilation and Sharing. For each dead bird or bat discovered at a wind farm, the information should be systematically recorded on a data sheet, noting the key information including the date, species, sex and age (when evident), observer name, turbine number, distance and direction from turbine, habitat surrounding carcass, condition of carcass (entire, partial, scavenged), and estimated time of death (for example, ≤1 day, 2 days). It is important for the data collected to be compiled in a readily understandable format, and, as feasible, to be publicly disclosed in the most appropriate manner. The data should be presented in a way that facilitates answering such basic questions as: (i) bird and bat mortality, total and by species, per turbine and per MW; (ii) mortality for species of conservation interest; and (iii) changes in mortality due to different wind farm management regimes (such as higher turbine cut-in speeds). Public disclosure of bird and bat monitoring data is needed for advancing scientific knowledge about the impacts of wind farms on birds and bats (and how to optimally mitigate them) in general, and for assessing plans for scaling up wind power within the same WRA in particular (such as Mexico's Isthmus of Tehuantepec). The dissemination of bird and bat monitoring data will be most helpful from a planning, adaptive management, and scientific standpoint if the data are explicitly and collaboratively shared with other wind developers, regulatory agencies, and international research networks and partnerships such as the Bats and Wind Energy Cooperative (www.batsandwind.org) and the Grassland and Shrub-Steppe Species Cooperative (www.nwcc.org).

OPERATIONAL CURTAILMENT

Operational curtailment refers to selected, short-term periods when the rotor blades are intentionally kept from rotating, in order to prevent turbine-related mortality during high-risk periods for birds or bats. Rotor blades rotate and produce electricity when they are aligned perpendicular to the direction of the wind. The blades are kept from rotating when they are aligned parallel to the wind, a position known as "feathering" (*posicion bandera* in Spanish). When the rotor blades are feathered, the turbines are not generating electricity. Even without operational curtailment for bird or bat conservation reasons, wind turbine rotor blades are often feathered because: (i) wind speeds are too low for the rotors to turn; (ii) wind speeds are too high, risking damage to the turbines if the rotors were kept turning; or (iii) for equipment maintenance or repair.

Increased Cut-in Speeds

Higher Cut-in Speeds Reduce Bat Mortality. The cut-in speed is the lowest wind speed at which the rotor blades are spinning and generating electricity for the grid. On large, modern turbines, the cut-in speed is typically 2.5-4.0 meters per second (m/sec), equivalent to about 9-15 kilometers per hour (kph) or 6-10 miles per hour (mph). Above a maximum safe wind speed for turbine operation—typically about 25 m/s (90 kph or 55 mph)—the rotors are programmed to stop turning (by feathering the blades) to avoid possible damage to the equipment. In this regard, bats are known to suppress their activity during periods of rain, low temperatures, and strong winds (Eckert 1982, Erickson and West 2002). Through the use of acoustic detectors and thermal imaging, bat activity in the U.S. has been found to be consistently higher during low wind periods (Arnett et al. 2006; Arnett, Huso et al. 2007; Reynolds 2006; Brinkman et al. 2006; Horn et al. 2008a). Since bats—whether feeding or migrating—tend to fly around wind turbines mainly at low wind speeds, increasing the cut-in speed can reduce bat mortality because it reduces turbine operating time during low wind speeds. At this writing, only three studies have been conducted worldwide on changing the cut-in speed—in Pennsylvania, USA; Alberta, Canada; and Germany. All three studies show remarkable results: Increasing the cut-in speed from the usual 3.5-4 m/s to about 6 m/s (at night and during the warmer months of year) reduced bat mortality by 50-74 percent, while reducing power generation by as little as 1 percent or less (Table 3.1). Such remarkable findings are explained by both biology and physics: bats fly around mostly at low wind speeds and primarily at night (and, at higher latitudes, during only a portion of the year), while low wind speeds in many WRAs are frequent, but contain relatively little energy that can be converted into electricity.

Further experimentation with cut-in speeds is clearly warranted, especially to find out what cut-in speeds would be needed to reduce bat mortality even further (for example, by 95 or 99 percent), and what the tradeoff with power generation and revenues would then be. Interestingly, in the Alberta study, the opportunity cost of power generation could have been further reduced (by more than half), if the turbines in use there could have been conveniently programmed to increase their cut-in speed only at night (when bats fly), rather than for 24 hours a day (Baerwald et al. 2009). In temperate latitudes, raising the cut-in speed may be particularly cost-effective because bats are present

Table 3.1: Effects of Increased Turbine Cut-in Speed on Bat Mortality and Power Generation

Wind Farm Location (Reference)	Casselman, Pennsylvania, USA (Arnett et al. 2010)	Alberta, Canada (Baerwald et al. 2009)	Germany (O. Behr, University of Hanover, unpublished data)
Regular Cut-in Speed	3.5 m/sec	4.0 m/sec	4.0 m/sec
Increased Cut-in Speed	5.0 m/sec and 6.5 m/sec (two treatments)	5.5 m/sec	5.5 m/sec
Bat Mortality Reduction	74% (combined data from both treatments)	58%	50%
Power Generation Reduction	0.3% for 5.0 m/sec; 1% for 6.5 m/sec	Cost less than 1% of annual revenue	Data not available

Source: Authors' compilation.

Photo: Edward Arnett, Bat Conservation International

A turbine at the Casselman, Pennsylvania wind farm, in a feathered state (blades pitched parallel to the wind) during operational curtailment to reduce bat mortality.

during only a portion of the year. Also, night-time wind speeds in many areas tend to be lower than during the day. Data from existing wind farms in North America and Europe indicate that a substantial portion of bat mortality occurs during the relatively brief summer-fall bat migration period (Arnett et al. 2008; Kunz, Arnett, Erickson et al. 2007; Dürr and Bach 2004; Johnson 2005; Brinkman et al. 2006; Fleming and Eby 2003; Cryan 2003). Even at tropical locations, bat abundance (and vulnerability to turbine kills) may be seasonal. For example, at La Venta II in Mexico (Appendix A), 42 percent of all bat mortality observed in 2008 was during the month of August.

Bat Conservation Measures. These findings indicate that an increase in turbine cut-in speed may become a key tool in substantially reducing bat mortality at wind turbines, while minimizing power generation losses. As noted in the section "Impacts on Bats," bats occur—and, to some extent, are likely to be killed—at most land-based wind farm locations; the main exceptions appear to be sites that are exceptionally arid or cold on a year-round basis. Accordingly, wind power projects could become significantly more bat-friendly by adopting one of the following two approaches:

- **Experiment with Cut-in Speeds and Monitor the Results.** Wind power projects—including those in operation today as well as the new ones—can significantly advance bat conservation and scientific knowledge by conducting relatively low-cost research trials with different turbine cut-in speeds and monitoring the effects on bat mortality. Arnett et al. 2009 describe the research methodology used at the Casselman, Pennsylvania wind farm. In this regard, technical assistance related to designing simplified versions of such research trials (in any country) is readily available from the U.S.-based Bats and Wind Energy Cooperative (BWEC, www.batsandwind.org), a collaborative program between Bat Conservation International, American Wind Energy Association, U.S. Fish and Wildlife Service, and National Renewable Energy Laboratory. Based on the research trial findings, the wind farm operators would then ideally choose an optimum cut-in speed that takes into account the implications for reduced bat mortality, as well as power generation. At the Uruguay Wind Farm Project, the Bird and Bat Monitoring Plan agreed between the utility UTE and the World Bank provides for: (i) operating the turbines at the standard cut-in speed of 4.0 m/s during Year 1; (ii) experimenting with 6.0 m/s (1/2 hour before sunset until sunrise) during Year 2, if the Year 1 monitoring finds (with correction factors) more than 5 dead bats/MW/year; and (iii) if bat fatalities drop significantly during Year 2, then continuing with 6.0 m/s during Year 3 (Rodriguez et al. 2009).
- **Presumptively Operate at Higher Cut-in Speeds.** Those wind farm operators who (for whatever reasons) do not conduct research trials with different cut-in speeds could nonetheless benefit bat conservation by choosing to operate at a slightly higher cut-in speed, on the presumption that, based on the strong scientific evidence to date, bat mortality will be significantly reduced. This would imply, as a first approximation, going from the "standard design" cut-in speed of 3-4 m/sec to around 6 (or at least 5) m/sec during the time of night (1/2 hour before sunset until sunrise) that most bats are flying.

Effects on Birds. Although changing the cut-in speed appears to be a critically important variable in reducing bat mortality during wind farm operation, research to date suggests that it may be a less important tool in substantially reducing bird mortality. The reasons for this difference appear to be that: (i) migrating birds often fly during high tail-winds and (ii) raptors, which tend to be especially vulnerable to wind turbine collisions, often continue to fly in search of food during high winds. For example, Hoover and Morrison (2005) noted that a disproportionate number of turbine-related Red-tailed Hawk mortality at Altamont Pass occurred on slopes where 90 percent of the kiting (hovering in place while searching for prey) took place. Kiting behavior occurs primarily during high winds that disrupt thermals and prevent effective soaring. Therefore, under these conditions, operational curtailment during low wind periods would reduce hawk mortality to only a limited degree.

Short-Term Shutdowns

Short-term shutdowns of wind turbines can be a valuable tool in reducing mortality during periods of especially high migratory bird or bat use. Short-term shutdowns can be some combination of: (i) seasonal, such as during peak migration periods, which (for species of special concern) might last for only a few days or weeks; (ii) time of day; (iii) on-demand in real time, such as when large at-risk flocks are spotted (as at the Mexico La Venta II project); or (iv) after a maximum "kill quota" is reached per turbine (based on independent post-construction monitoring), as has recently been incorporated within the environmental permit for a new wind farm in Highland County, Virginia, USA (TNC 2008). During short-term shutdowns, the rotor blades are feathered. Short-term shutdowns are most cost-effective when the bird or bat species of concern are migratory in some sense, in that they spend a relatively small portion of the year in the wind farm area. However, shutdowns are harder to justify as a wildlife mitigation tool when the species of concern spend(s) a large proportion of the year around the wind farm, as is the case with White-tailed Eagles at Smola Island and various raptor species at Altamont Pass.

The World Bank–supported La Venta II project in Mexico's Isthmus of Tehuantepec shows that short-term shutdowns are feasible and effective in preventing large-scale mortality of migratory birds. For that project, located within a world-class bird migration corridor, the Comisión Federal de Electricidad (CFE), the wind farm operator, and INECOL, the independent monitor, have agreed upon a specific procedure for real-time, short-term shutdowns (feathering) of some or all turbines when large flocks of birds are detected, by visual observation and/or radar, flying towards the wind farm near RSA height. Since they began work in the autumn of 2007 until at least 2009, the INECOL monitors have requested brief shutdowns, lasting less than half an hour, and then only infrequently—about three times per year (R. Villegas-Patraca *pers. comm.* 2009). This is because except during unusual weather conditions, the large migratory flocks tend to fly well above the top RSA height (72 m). Thus, the procedure for short-term turbine shutdowns at La Venta II remains available as an infrequently used, but important contingency measure that has only a negligible impact on power generation or revenues. Short-term shutdowns during peak spring and fall migration periods, using a MERLIN avian profiling radar system, are also planned for the new Penascal (by Iberdrola Renewables) and Gulf Wind (by Babcock & Brown) projects along the Texas coast (USFWS 2010).

For bats, changing turbine cut-in speed will often be the most cost-effective curtailment option because most bat activity and turbine-related mortality occurs during low wind periods. However, in those cases where large numbers of migratory bats are known to occupy a site consistently during relatively predictable times,[3] scheduled shutdowns could be even more effective than changing cut-in speeds.

Financial and Economic Considerations

The implementation of operational curtailment (increased cut-in speed or short-term shutdowns) should take into account the effects on the financial returns of the wind power operator. Because many regulatory systems only pay wind generators for the electricity they generate, rather than also for capacity as is the case for "firm" power generators, a wind project—like a run-of-river hydropower plant—might be disproportionately affected by measures that curtail power production. In addition, many wind power facilities have been designed (without bats in mind) to operate at low wind speeds in order to maximize the capacity utilization factor as well as power output. Any possible change in maintenance or other operating costs involved with a particular operational curtailment regime also needs to be considered. It is therefore important for wind project planners to assess in advance the financial and power supply implications of any proposed operational curtailment regime, in comparison with the potential reductions in bird or bat collisions. Such cost-benefit assessments—carried out as part of project feasibility and environmental assessment studies—could help wind power developers and public regulators to: (i) understand the full implications of each proposed operational curtailment measure under expected operating scenarios and (ii) define the relative effectiveness of different alternatives for reducing bat or bird mortality.

Wind Farm Maintenance Practices

While sometimes overlooked by project planners, wind farm maintenance practices are often an important tool in managing biodiversity and related environmental impacts. The practices that merit attention include equipment maintenance and landscape management at wind farms.

Equipment Maintenance. In some cases, proper equipment maintenance can prevent unnecessary bird mortality or other environmental damage. A case in point is the Mexico La Venta II wind farm, where some of the wind turbine nacelles had round holes at their base with the covers missing. Because these holes were uncovered, American Kestrels were using them for shelter (and probable nesting), but with fatal consequences because of the close proximity of the rotors; five kestrel carcasses had been found under four such turbines. Interestingly, one of the five turbines in the Uruguay Wind Farm project had similar round holes with the caps missing; American Kestrels also occur there (although no carcasses had been encountered). At both projects, on-site personnel agreed to cap the holes. This experience suggests that examining turbine nacelles for unnecessary, open holes that could attract various species of birds or bats to enter (at their great peril) would ideally be a standard part of wind turbine maintenance checklists.

Rules of Good Conduct. As during construction, it is important to instruct wind farm employees and contractors to follow rules of good environmental and social conduct during project operation. Among others, such rules include prohibitions on off-road driving (which damages vegetation and compacts soils), speeding (risky to people, livestock, and wildlife), wildlife capture, bush meat purchase, firearms possession (except by security personnel), and inappropriate interactions with local people. These simple rules can effectively prevent or reduce needless environmental damage, and will help to maintain the wind industry's overall image as a "good citizen." The absence of, or failure by wind farm managers to enforce, such rules can cause unnecessary environmental and social problems. For example, in the Huitengxile area of Inner Mongolia, China, careless off-road driving by wind farm personnel has degraded extensive areas of grassland with naturally sparse vegetation, with adverse consequences for the livestock of native herders (R. Spencer *pers. comm.* 2009).

Landscape Management at Wind Farms. Landscape management decisions are ideally discussed and coordinated between landowners, area residents, wind farm operators, environmental agencies, and other relevant stakeholders. Landscape management at wind farms needs to take into account a variety of different—and sometimes conflicting—objectives, including those discussed below.

- **Maintaining Preexisting Land Uses.** One of the main advantages and "selling points" of wind power, in comparison with most other types of (renewable and non-renewable) energy development, is that preexisting land uses such as crop cultivation and grazing can often continue more or less unimpeded. To the extent that preexisting land uses are constrained, such as by the footprint of the wind turbine platforms and interconnecting roads, the landowners or other affected people need to be appropriately compensated (see Chapter 4).

- **Conserving and Restoring Natural Habitats.** From a biodiversity conservation standpoint, wind farms are ideally situated on agricultural lands or other previously disturbed areas, rather than intact natural habitats. This is because, aside from the bird and bat collision risks, the wind turbine platforms and access roads will invariably fragment any natural habitats upon which they are sited. However, those wind farms that are situated within natural habitats might provide a valuable conservation opportunity, if the remaining natural vegetation (other than what is needed for turbine platforms and roads) is intentionally preserved as part of the wind project's environmental permit or landowner agreement. For example, the low-growing native thorn forest of the southern Isthmus of Tehuantepec in Oaxaca, Mexico is the world's only habitat for the endemic Cinnamon-tailed Sparrow; it also harbors other animal and plant species of conservation interest. None of this habitat is presently within any type of protected area. Consequently, wind power projects located within this WRA could make a very positive contribution to biodiversity conservation by working out agreements with landowners to maintain intact this native vegetation, and adjusting accordingly the amounts paid to landowners for hosting the wind power facilities. Whenever feasible, wind farm land that was previously disturbed should be restored with native vegetation, if it is no longer needed for agriculture or other intensive uses. When restoring degraded land on wind farms, it is important (as elsewhere) to avoid planting non-native, invasive species that can spread widely and cause considerable ecological and economic damage.

- **Managing Land for Species of Conservation Interest.** As part of a wind project, agreements could be reached with landowners to ensure that the land comprising the wind farm is managed optimally for species of conservation interest. For example, some North American grassland bird species are in decline because meadows and hayfields are being mowed too early in the summer; this practice destroys active nests and prevents successful reproduction (Lebbin et al. 2010). Some of these same species, such as Bobolink, are sometimes killed by wind turbines because breeding males perform aerial territorial displays, often flying at RSA height. One way to help offset the turbine-related mortality of breeding males would be to secure agreements with landowners around wind farms to adjust their hay mowing schedules towards later in the summer, after the ground-nesting birds have fledged, thereby promoting improved breeding success. This approach has been proposed by Iberdrola Renewables in their Avian and Bat Protection Plan (Iberdrola 2008).

- **Deterring Bird or Bat Use.** Landscape management at wind farms can sometimes be a useful tool for reducing bird mortality, by making the habitat less attractive to the species at significant risk of colliding with the turbines. For example, the Altamont Pass WRA has especially high raptor mortality (Box 3.2), in large measure because the grassy hills, heavily grazed by cattle, support high densities of California Ground Squirrels *Spermophilus beecheyi*, Desert Cottontails *Sylvilagus audubonii*, and other small mammals that attract large numbers of hawks, eagles, and owls. Past efforts to reduce raptor mortality by poisoning these small mammals have generally proved ineffective (Smallwood and Thelander 2004). Moreover, the use of poisons and pesticides involves many other environmental and human health risks, particularly when the storage, proper use, and disposal of these compounds are not rigorously controlled. In this regard, any proposed habitat modification measures at wind farms should be assessed very carefully and, where possible, attempted initially on only a small scale, due to the potentially adverse impacts upon other species (besides those affected by turbine mortality). Wind farms can be made less attractive to vultures and other scavenging birds through prompt removal of dead cattle or other carrion, as is required at the Mexico La Venta II project. In the case of bats, the "edge effect" provided by wind farm clearings in forested areas is believed to attract bats (in addition to the bats' attraction to rotating turbine blades, discussed in Box 3.4). However, habitat management measures for discouraging bat activity around wind turbines have rarely, if ever, been used or proposed.

- **Facilitating Bird and Bat Monitoring.** Post-construction monitoring of bird and bat mortality is most accurate when the area around each turbine (with a radius as long as the turbine's maximum RSA height) has little or no vegetation since searcher efficiency is increased. This typically poses a tradeoff with the objective of maintaining or restoring natural vegetation, in order to minimize each turbine's land footprint. Conserving or restoring the natural vegetation is an especially high priority in those cases where the vegetation is needed for biodiversity conservation (as at the Mexico La Venta II project, where the native thorn forest harbors endemic species), erosion control, or other important environmental objectives. Otherwise, maintaining minimal or short ground cover is advisable around each turbine, for as long as post-construction monitoring involving searches for dead birds and bats will be taking place.

Prior Planning of Vegetation Management. For optimum results, vegetation management at wind farms should not be an afterthought, but rather, should be deliberately planned and recorded within the project's EMP. As an example, the effectiveness of post-construction monitoring at the Casselman, Pennsylvania wind farm is constrained because the land around certain turbines cannot be cleared, due to prior, binding land use commitments made by the landowners. With careful prior planning, alternative sites for specific turbines within the same overall wind farm location could have been chosen that would have allowed for the land right around each turbine to be cleared, thereby facilitating more accurate bat mortality estimates.

Managing Public Access to Wind Power Facilities. Decisions about how much public access to allow at wind farms and associated transmission line corridors involve balancing a variety of environmental and social objectives. These objectives include, among others: (i) maintaining previous land uses; (ii) ensuring public safety; (iii) minimizing the risk of sabotage or theft of wind power equipment; (iv) protecting threatened species and ecosystems; and (v) promoting local tourism and recreation. In areas where hunting is inadequately regulated from a conservation standpoint or illegal hunting is common, a wind project can substantially mitigate or enhance its biodiversity impact by effectively enforcing a prohibition on hunting and shooting within the wind farm area. Such a prohibition is highly advisable in any case to reduce the risk of damage to turbines from gunshots; it could also help to offset turbine-related bird mortality by reducing the hunting-related mortality. At the Mexico La Venta II project, anecdotal evidence suggests that local populations of small mammals and some bird species have increased because hunting by local residents within the wind farm area has been curtailed. However, in situations where new restrictions on formerly legal hunting activities could lead to adverse impacts on local livelihoods, it is good practice for the project developer to take this into account and to design and implement remedial measures as needed.

Conservation Offsets

Conservation offsets (sometimes called compensatory mitigation) can be a very useful tool for mitigating the biodiversity-related harm and enhancing the overall conservation outcomes of a wind power project.[4] Off-site conservation investments can successfully mitigate the biodiversity-related impacts of wind projects, in a complementary manner to the on-site measures within the wind farms' boundaries. To be successful, off-site conservation activities need to be properly planned and executed, with clear implementation responsibilities and adequate, up-front funding commitments as part of the overall wind power project. Off-site conservation options include:

- **Habitat Conservation.** To offset some degree of inevitable damage to natural habitats or species of conservation concern, natural habitats of similar or greater conservation value to what is affected by the wind power project can be protected and managed, with financial support from the wind project sponsors. For example, to offset the displacement of Lesser Prairie Chickens *Tympanuchus pallidicinctus* from their natural grassland habitat by a wind power project in Oklahoma, USA, the project sponsor (Oklahoma Gas and Electric) has agreed to provide funding to the Oklahoma Department of Wildlife Conservation for long-term habitat protection and improvement elsewhere in northwestern Oklahoma (USFWS 2010). Other options for off-site habitat conservation can include support by wind project sponsors for the establishment or strengthening of specific protected areas, reforestation or other habitat restoration or enhancement of key sites, or payment into a national or other protected areas fund (in those countries where such a fund already exists).

Species Management. The incidental killing of species of conservation interest by wind power facilities can sometimes be adequately offset by efforts to enhance the species' populations in other ways. For example, a useful offset to the anticipated future mortality at wind farms of the endangered Indiana Bat (as wind power in the northeastern United States continues to expand) could be specific measures to increase the protection and management of the hibernation caves used by this species. For migratory ducks, geese, and other waterfowl that are legally hunted as game birds, any significant mortality from wind farms or associated transmission lines could readily be offset by reductions in annual hunting quotas through shorter hunting seasons or reduced bag limits, assuming there is: (i) adequate information about the extent of wind power-related mortality and (ii) adequate political will, so as not to interfere with science-based adjustments to hunting quotas.

Notes

1. The World Bank's Natural Habitats Policy (OP 4.04, Appendix A) contains a detailed definition of "critical natural habitats"; the IFC's Performance Standard 6 on Biodiversity Conservation and Sustainable Natural Resource Management has a broadly similar definition of "critical habitat."
2. Echolocation involves determining the location of something by measuring the time it takes for an echo to return from it. Most bat species use echolocation to navigate and maneuver while flying in the dark, successfully avoiding collisions with most objects except (for unknown reasons) wind turbines.
3. See Cryan and Brown 2007 for an example of migrating Hoary Bats.
4. Under the World Bank's Natural Habitats policy OP 4.04 and Forests policy OP 4.36, as well as the IFC's Performance Standard 6 on Biodiversity Conservation and Sustainable Natural Resource Management, some type of conservation offset would be needed in certain cases where a wind project causes significant loss or degradation of natural habitats. Such loss or degradation could happen if, for example, a new wind farm were to (i) clear away a significant area of biologically distinctive ridge-top vegetation for the installation of turbines and access roads; (ii) fragment a significant area of natural forest or other native vegetation, through access roads and transmission lines; or (iii) ecologically degrade an area by removing a significant proportion of its bat, raptor, or other bird population. From an ecological standpoint, a terrestrial natural habitat comprises not only the land surface and vegetation, but also the airspace above, as high up as there is significant natural biological activity.

CHAPTER 4

Addressing the Social Impacts of Wind Power

Typical Social Impacts in Wind Power Development

Introduction

Visual representations of wind farms tend to show an array of massive, lonely turbines on a grassy plain, along a high ridge-top, or on the surface of the ocean. Like the installations associated with any power generation technology, however, wind turbines do not exist in a social vacuum; as human creations they cannot be separated from the social and cultural settings in which they are designed, built, and operated. The first part of this chapter looks at some of the key socio-cultural and economic impacts of wind power development, which may be negative or positive. It begins by considering a set of negative impacts that are associated with how wind projects, with an emphasis on wind turbine facilities, are perceived by the people living in their vicinity, and which are sometimes considered to fall under the rubric of "local environmental impacts" in environmental assessment (EA) parlance. The discussion continues with a review of positive and negative livelihood impacts, with benefits such as employment considered as an example of the former, and losses such as those tied to land acquisition for wind farms considered as an example of the latter. Finally, the chapter concludes with an examination of the special circumstances surrounding land tenure insecurity in wind resource areas, wind power development on indigenous lands, impacts on physical cultural resources (PCR) uncovered in the course of wind farm construction, and offshore wind power development.

Local Nuisance Impacts

VISUAL IMPACTS

Even though old-fashioned windmills are widely regarded as attractive and "quaint," large modern windmills are sometimes considered to be an eyesore. Modern wind turbines that generate grid-based electricity are inherently large, which makes them visible over long distances. While the largest wind turbines can be seen from as far as 30 km away, the most significant visual impacts are likely to occur within 5 km of a wind farm. A special type of visual impact at this close range is shadow flicker, which is described below.

To date, visual impacts have emerged as a leading socio-environmental constraint to installing new wind farms and associated transmission lines, especially in North America and Europe. Government authorities have often given concessions for these installations without considering how they will be regarded visually by the people living nearby. In this regard, prior consultation is important in term of minimizing conflicts with local stakeholders. In the developing world, visual impact tends to be less of a factor, especially in remote areas where wind turbines are seen as a novelty or sign of progress. However, visual impacts are definitely of concern in more affluent areas of developing countries (for example, Uruguay). Among other examples, as the experience in the Caracoles case (Appendix C) has shown, special care is needed in windy areas of high tourism potential.

Public Attitudes Towards Wind Power. A Danish survey (Gipe 1995) found that those in favor of renewable energy sources in general, and wind power in particular, were more positive about the impacts associated with the installation of turbines, which they found less visually intrusive and less noisy.[1] Other studies (for example, Wolsink 1996) show that in areas where there was significant public resistance to wind projects, the people involved were not reacting against the turbines, but rather against the outside agents who wanted to build them. Often local people are excluded from decision-making related to site selection and other details concerning wind projects, which may engender resentment toward the project developer. The "not-in-my-backyard" (NIMBY) phenomenon has been shown to also play a role in shaping public attitudes towards wind power in some areas (Krohn and Damborg 1999).

Careful site selection and design modifications can significantly reduce visual impacts of wind power projects. For example, at the Sierra de los Caracoles wind farm in southern Uruguay, the project developer paid special attention to how the installation of windmills on the highest point of an otherwise flat area—a ridge line of the Sierra de Caracoles mountain range—would be received by the largely affluent owners of cattle ranching estates and vacation homes in the vicinity. Instead of approving the original design, which would have erected 12 relatively short towers (each with an 800 kW generation capacity) on the ridge, the developer adopted a modified design consisting of five taller towers with a 2 MW generation capacity apiece. The new configuration occupied less space along the ridge and, despite the added height, was perceived to have a smaller aggregate visual impact. It had the added benefit of reducing the project's impact on the most prominent example of cultural patrimony in the area, an old stone wall spanning the ridge top that was built in the seventeenth century as a dividing line between estates.

Visual impacts may be particularly sensitive in areas with high tourism potential and in popular recreation areas. A case in point is the site of the proposed Cape Wind project is in an offshore area close to Cape Cod and the islands of Nantucket and Martha's Vineyard (USA), which are popular summer destinations. Visual impacts were one of the factors cited in the initial opposition to the project although these seem to have been superseded by other issues (potential bird mortality, navigation and fisheries impacts, and recreational impacts). Among the various modifications made in response to public concerns, the Cape Wind sponsor agreed to paint the windmill towers a mixture of light gray and light blue, which would allow them to blend in with the horizon (Cape Wind Associates, *pers. comm.* 2010).

Shadow Flicker. Wind turbines, like other tall structures, cast a shadow on the surrounding area when the sun is visible. For people who live close to the turbine, it may be annoying if the rotor blades cause a flickering (blinking) effect while the rotor is in motion. (Shadows, of course, are longer around sunrise and sunset, and shorter at noon.) Shadow casting is generally not regulated explicitly by planning authorities. However, in Germany, there was a court case in which the judge allowed 30 hours of actual shadow flicker per year for one individual's property, with the time limit applying only during the hours when the property was in use (Danish Wind Industry Association 2003).

Estimating the exact shape, place, and time of the shadow from a wind turbine requires computation, but professional wind software programs can do this very accurately, even in hilly terrain, and with house windows of any size, shape, location, and inclination facing in any direction. Shadow geometry varies by latitude, and the length of the shadow is much more a function of rotor diameter than of hub height. For more information, see the *EHS Guidelines* (World Bank Group 2007).

Noise from Wind Turbines

Types of Noise. The noises produced by the operation of wind farms include "turbine hum" and "rotor swish." The first is emitted from the machinery in the nacelle of a turbine. The second is produced by the friction between the wing blades and the wind. Noise levels from within an onshore wind farm are mostly in the range of 35-45 dB(A) (decibels corrected or "A-weighted" for sensitivity of the human ear). These are relatively low noise levels compared with other common sources such as ambient nighttime noises in the countryside (about 20 to 40 dB[A]) or the noises of a busy office (about 60 dB[A]).

The Role of Individual Perception. The exact level of sound from a given wind farm can be measured objectively. Nevertheless, the impact of a given noise level is often a function of individual perception. Some people might perceive relatively low levels of sound from the turbines as unacceptable noise, while other persons might perceive the same level of sound as acceptable. The attitude toward wind power of the two persons would, everything else considered equal, be different even though they are exposed to the same level of disturbance.

The subjective judgment of the level of noise from a wind farm has been shown to depend on individual characteristics such as a person's attitude towards wind power production, his or her age, gender, and so forth (Manwell et al. 2002, Pedersen 2004, Wolsink and Sprengers 1993). For example, a survey in Denmark found that people with no specific experiences with wind power believed that noise associated with it is louder than did those who actually live beside turbines. Also, men believed that turbines are noisier than women do (Krohn and Damborg 1999). Some surveys have shown that middle-aged people tend to be more critical of the noise impact of wind power compared to other age groups.

Setback Options for Dwellings. Although noise is commonly cited as a source of public concern about proposed new wind farms, the problem is effectively mitigated by ensuring that wind turbines are placed at least 130 meters from a dwelling or other area where people typically congregate. This approach is demonstrated in Figure 4.1 (see the Color Photo Insert at the end of this paper). However, to avoid the hazards of ice throw from rotor blades in cold climates, or the very small risk of blade throw from a broken turbine, a setback of about 300 m is typically recommended (World Bank Group 2007), although some private developers have favored adopting a setback as large as 500 m (J. Villegas, *pers. comm.*). It is important to note that there could be a significant difference between the level of noise expected ex-ante by people close to the site, and the actual level of noise perceived once the plant is in operation. This is a relevant factor to be considered during the planning stages of a project.

Noise is often perceived as a relatively minor problem in wind power development today, even when set-back restrictions are not observed. Sound emissions of new turbine designs tend to cluster around the same values. Moreover, it is usually easy to predict the sound effect from wind turbines in advance. Because of the difficulties inherent in distinguishing turbine hum or rotor swish from natural background noise, making such calculations ex-ante is generally preferred to trying to measure the sounds of an operating wind farm. At the same time, calculating potential sound emission from wind turbines is generally important in order to obtain permission from the public planning authorities to install wind turbines in densely populated areas. Therefore, in most places in the world, public authorities rely on calculations rather than measurements when granting planning permission for wind turbines.

Other Nuisance and Safety Issues

Telecommunications Interference. Operating wind turbines tend to interfere with the signals received by radar and telecommunications systems, including aviation radar, radio, television, and microwave transmission. These impacts are likely to be significant when the wind turbines are within the line-of-sight of the radar or telecommunications facility (see *EHS Guidelines* for further detail).

Aviation Safety. Wind turbines can pose a risk of aircraft collisions if they are located too close to airport runways. In addition, in agricultural areas, the presence of wind turbines will preclude aerial spraying of pesticides in the immediate vicinity, due to the risk of airplane collisions. In view of the environmental risks posed by many pesticides, such restrictions on aerial spraying could be viewed as an environmentally beneficial side effect.

Blade and Ice Throw. There is a very small risk of a loose rotor blade being thrown as a result of severe mechanical failure; any such loose rotor blade would almost certainly fall within 300 m of the turbine. In cold climates, there is also a small risk of rotor blades throwing off chunks of accumulated ice when they begin to rotate; most ice pieces fall within 100 m of the turbine (see *EHS Guidelines*). Some wind turbine rotor blades (as with aircraft propellers) are equipped with heating elements to reduce ice formation.

Socioeconomic and Cultural Impacts

Overview. There are a number of examples of beneficial socioeconomic impacts of wind power, even for populations living in the immediate vicinity of wind projects and their ancillary infrastructure. Still, a number of complex and site-specific issues may arise:

- Concerning the maintenance and/or enhancement of people's livelihoods, the impacts tend to be positive. The nature of the footprint of wind power projects permits preexisting land uses to continue. When direct economic benefits are factored in, such projects can increase income for rural landowners in the wind farm area and help boost local economies. Still, the sum of positive benefits—particularly those at the local level—can vary widely from project to project as detailed below.
- In most regions, the negotiation of leases or easements is the most common way of acquiring land for wind farms. This represents a major advantage over other types of renewable and non-renewable energy technologies that depend

on formal expropriation, especially when it requires large-scale compensation and/or the replacement of lost land and other assets as in the case of many hydroelectric dams.

■   Besides impacts related to visual aspects and noise, wind farms can raise sensitive issues concerning historic and archeological sites and cultural values. At offshore sites, additional issues include impacts on fisheries and navigation.

Cultural Impacts. When considering the potential for scaling-up wind power across the globe, and particularly in developing countries, it is important to take the cultural impacts involved into account. While this is related to considering the cultural dimensions of wind power—another aim of this paper—it is also distinct in that it involves examining how the beliefs, preferences, and behaviors of often poor, marginalized people are shaped by the introduction or expansion of wind power in the areas where they live and work.[2] They are often poor and marginalized because, in many parts of the developing world and certainly in Latin America and the Caribbean (LAC), the areas with the highest wind power potential are remote deserts, plains, and mountaintops—precisely those places where lower-income rural populations tend to be found. The possibility also arises that such places will overlap with, or be contained entirely within, the territories of indigenous peoples.

The Potential for Cultural Disruption. In cases where wind power development is externally driven and where there are significant differences in key characteristics—such as ethnicity, race, or caste—of the local population versus those of the people sponsoring a given project, the introduction of the wind project could prove culturally disruptive to the former population. In this regard, the project's introduction could unleash forces of cultural change—such as the conversion from a local economy based largely on barter to one based on cash transactions—that those living in the project influence area might find undesirable, unwelcome, and even harmful. Of course, the decision to develop is hardly a one-way process; through their own attempts to shape project plans and actions, participating indigenous community members can leave a distinct stamp on how—or even whether—the project in question is carried out. Issues associated with the management of wind power projects in socially and culturally diverse and/or sensitive settings are discussed further below.

LIVELIHOOD IMPACTS

Wind turbine platforms and access roads have a relatively small land footprint, so most preexisting rural land uses—including crop cultivation and livestock grazing—can continue undisturbed. For transmission line rights-of-way (ROW), most land uses that do not involve houses or tall crops are compatible. There is thus considerably less livelihood loss or physical displacement than with other types of renewable power generation such as hydroelectric reservoirs or solar farms, making wind more attractive from the point of view of minimizing adverse social impacts. Indeed, at the local level, wind power has shown great potential for generating a number of economic and social benefits, which are discussed below.

Overall Economic Benefits. Wind farms potentially offer rural landowners and communities a variety of economic benefits, notably:

- ■ **Local employment.** Wind power can boost the local economy by creating short-term jobs during the construction phase and some long-term employment during the operation and maintenance phase.
- ■ **Increased income.** Wind farm leases offer rural landowners significant additional income, even though most of the leased land remains available for continued farming or ranching.
- ■ **Income diversification.** Wind power offers a constant source of income every year, in contrast to income from ranching or agricultural activities, which can vary from year to year and, in the case of agriculture, is forthcoming mainly at harvest time.

Photo: China Wind Power
Wind farms enable most preexisting land uses, such as livestock grazing, to continue in the spaces between each turbine.

Employment Benefits. Job creation for wind power projects can be significant, although there is considerable variation in estimates due to the range of calculation methods and assumptions used. As a class, wind power projects tend to be less labor intensive than infrastructure projects in other sectors such as transport or water and sanitation. In the United States, for example, it has been estimated that wind power creates about 3,400 jobs per US$1 billion of disbursements, compared to 23,000 jobs per $1 billion disbursed for rural electrification in Peru, or 66,667 jobs per $1 billion disbursed for the expansion of water networks in Honduras (Tuck et al. 2009). Table 4.1 provides detailed figures regarding the employment benefits generated by different kinds of projects in the energy, transport, and water and sanitation sectors in both the United States and Latin America.

Employment During the Construction Phase. In the United States, a 50 MW wind farm can generate about 40 full-time jobs during the construction phase. These construction jobs generally last about a year. In the case of longer-lasting operations and maintenance jobs, it is estimated that one job is created for every five to eight MW of installed capacity. A 50 MW installation would thus create between six and ten permanent positions, with higher levels possible in developing countries (Winrock International et al. 2003). In the case of the 20-turbine First Wind facility on the island of Maui, Hawaii

Table 4.1: Employment Generated by Development Projects in Three Sectors in Latin American Countries and the United States

	Qualified workers	Non-qualified workers	Domestic inputs (mainly material)	Foreign inputs (mainly equipment)	Others	Total	Annual Direct Employment (per US$1 Bil/yr)[*]
Transport							
Colombia—Access to neighborhoods (streets)	15%		49%	16%	14%	100%	22,500
Colombia—Feeder routes for Transmilenio		43%	27%	23%	6%	99%	35,833
Brazil—Roads	3%		22%	63%	3%	100%	16,577
Argentina—Rosario—highways	1.3%		60%	38%	0%	100%	1,650
Water and Sanitation							
Honduras—Improvement of water capture	28%		40%	20%		100%	43,833
Honduras—Rehabilitation of water networks	30%		40%	10%			58,333
Honduras—Expansion of water networks	20%		40%	10%		100%	66,667
Honduras—New Treatment Plant	10%		80%	0%		100%	25,000
Colombia—Expansion of WSS networks	8%		32%	4%		100%	100,000
Brazil—Rain Drainage networks	8%		48%	28%	0%	100%	34,001
Brazil—Sewage	4%		68%	17%	0%		21,746
Energy							
U.S.—Solar PV	3%-5%		95%-97%			100%	
U.S.—Wind Power	4%-6%		94%-96%			100%	
U.S.—Biomass	1%-2%		98%-99%			100%	
U.S.—Coal Fired	1%-2%		98%-99%			100%	
U.S.—Natural Gas-fired	2%-4%		96%-98%			100%	
Brazil—Hydropower	5%-10%		90%-95%			100%	
Peru—Rural Electrification	14%		7%	53%	0%	100%	23,000

Source: Reprinted from Tuck et al 2009.
[*]These estimates were based on an hourly wage of $3 for non-qualified workers and $6/hr for qualified ones for 2,000 working hours a year.

(USA), about 200 workers were hired during the construction phase, but there are only 7 full-time employees at the operational wind farm. Of these, 3 positions are for environmental mitigation, specifically endangered species monitoring and habitat restoration and enhancement (King 2009).

For wind projects in developing countries, employment levels are generally much higher due to lower labor costs. In India, for example, 10 to 15 people may be employed to operate and maintain a project with only a few turbines. By the same token, in areas with exceptional wind resources and the potential for large-scale wind energy development, there may be sufficient demand for the establishment of local facilities that produce wind turbine towers or other equipment. For example, it is estimated that in the southern Tehuantepec Isthmus region of Mexico, where there are a number of wind projects under development, a wind turbine tower manufacturing facility producing one hundred 65-75 meter towers a year would create about 100 jobs (Winrock International et al. 2003).

Employment in Project Operation and Maintenance. At the Sierra de los Caracoles site in southern Uruguay, local employment generated by the project was relatively weak. According to the project developer, state electricity utility UTE, about 39 percent of the total labor needed for the project during the construction phase came from the immediate area. Now that the wind farm is operating, UTE is employing one person from nearby San Carlos town as a security guard. Day-to-day operation of the wind farm is being handled by Uruguayan and foreign technicians with Eduinter, the Spanish firm that assembled the windmills. In addition, maintenance services are currently being provided by Vestas, the manufacturer of the wind turbine equipment for the wind farm. Thus, the promise of permanent employment in relation to scaling up wind power in the developing world may be more easily fulfilled by prequalified workers from important urban centers in host countries or from overseas.

Lease Income. Landowners in the immediate area where wind turbine facilities are constructed can benefit from the additional income provided by lease payments.[3] In theory, such payments are made in exchange for the right to exploit the wind blowing across the owners' lands. However, in practice, they are treated separately from compensation that would be payable for the expropriation of lands needed for turbine platforms, access roads, project offices, and other physical installations (see the following section), although in some cases they incorporate such compensation. Lease payments can be fixed or can take the form of royalties based on an agreed percentage of a wind project's anticipated gross revenues. In the latter case, payments vary depending on power sales prices and project capacity factors. For instance, it has been estimated that in Mexico's southern Tehuantepec Isthmus, royalty-based land lease payments—assuming power sales prices between $0.035 to $0.08 per kWh (kilowatt hour), a 40 percent capacity factor, and a contract paying 2 percent of gross revenues—would provide payments of about $320 to $728 per hectare (ha). In this particular area, capacity factors tend to remain fairly constant, making the price at which a particular project can sell energy the main variable (Winrock International et al. 2003).

In some cases, lease payments have been shown to result in dramatic increases in household incomes. At the La Venta II site in the Mexican state of Oaxaca, lease payments on average have doubled the annual income of local beneficiaries. In December 2008, when a World Bank team visited the site, the cooperative landowners (*ejidatarios*) generally expressed satisfaction with the fixed annual payments they had been receiving. Payments have been made on an annual basis, providing the *ejidatarios* with a reliable income stream that they can use to complement earnings from more traditional uses of their lands (sorghum cultivation and, to a lesser extent, cattle grazing). The attractiveness of the payments as a household-centered social benefit influenced a few *ejidatarios* within the borders of the area designated for the wind farm—who for various reasons initially opted not to participate in the project—to submit a formal petition to the Comisión Federal de Electricidad (CFE), the project sponsor, for inclusion. However, after evaluating the wind farm expansion needs implied by the petition, CFE decided that it did not have sufficient financial resources to accept it.

LAND ACQUISITION AND TENURE ISSUES

The overall amount of land required for a wind power project can vary significantly depending on wind and terrain conditions as well as on wind turbine size. In areas where the prevailing winds are unidirectional, turbines can be placed closer together in a row perpendicular to the wind direction; if the wind is multidirectional, greater spacing is required. The spacing between rows also needs to increase or decrease in proportion to the average wind speed (R. Gebhardt, *pers. comm.* 2010). Prospective wind farm sites with rolling hills or mountain ridges require greater spacing than flat terrain does. One study has estimated that, in the best-case scenario (that is, flat terrain with unidirectional winds), the minimum area needed is from approximately 7.7 to 10.4 ha/MW, which is equivalent to a minimum of 6.6 ha per wind turbine, depending on its specifications. In a site with rolling hills or mountain ridges, minimum area requirements increase to 15-20 ha/MW, while in a site with multidirectional winds, at least 20 ha/MW are needed for flat terrain, increasing to 38 ha/MW for rolling hills and more than 76 ha/MW for mountain ridges (Winrock International et al. 2003).[4]

Means of Acquiring Land

Three Main Options. Because of the relatively small direct footprint of wind turbine platforms and ancillary infrastructure, expropriation is often not required. Similarly, outright land purchase for wind farms is not widely used since it constitutes an additional up-front expense in a capital-intensive project. Rather, negotiated lease/rental arrangements or easements are much more common, whether in the case of privately or communally owned lands. Such leasing or easements tend to involve making payments for the use of the land needed for installation of the wind farm platform, building of access roads, and tending of transmission lines. They may also involve the sharing of a broader range of benefits. Below is a detailed discussion of the three possible means of acquiring land for wind power development, in the order of least to most common, and the impacts associated with each option.

Expropriation. Apart from the adverse social, economic, and environmental impacts that land expropriation can provoke, there are several reasons why it is so often eschewed. First, local residents can usually continue their normal income-earning activities while the wind turbines are spinning. Second, private companies may not have the legal authority or the experience needed to undertake the expropriation process, and yet increasingly, it is these companies—rather than public utilities—that are spearheading the spread of wind power across the developing world. Third, whether they are private or public, wind power developers may not have the time or staff resources needed to see a formal land expropriation process through, especially when (as is often the case in developing countries) the compensation offered by law is minimal. This, in turn, can lead to protracted negotiations with affected people over compensation amounts, or to legal battles that can take years to resolve. Fourth, wind developers tend not to have any interest in becoming involved in agriculture or ranching; they would rather just rent the lands needed for their wind schemes and allow the local residents to continue carrying out these activities themselves (Winrock International et al. 2003).

The Sierra de los Caracoles project in southern Uruguay provides an example of the use of expropriation for the installation of five windmills and the construction of a control center. The expropriation was undertaken by UTE, the project's public sector sponsor, via an executive decree issued by the Uruguayan government in August 2006. Sections of three contiguous plots totaling nearly 27 hectares were expropriated atop a hilly ridge where the wind farm is sited. Two relatively well-off landowners were affected; one of them expressed dissatisfaction with the compensation amount offered by the utility. When subsequent rounds of negotiation and reappraisal failed to produce an agreement between UTE and this landowner, the matter was referred to the Uruguayan courts, where it was finally resolved in mid-2009 to the satisfaction of all parties. Construction and operation of the Caracoles wind farm went ahead while the second landowner's petition was still being heard by the courts on the basis of a specific legal clause, but such dispositions in national law are more the exception than the rule.

Compensation Payment Standards. In the case of the Caracoles project, the compensation paid for both the expropriated land and the easements needed was roughly on par with the "replacement cost" standard established by OP 4.12. This is a method of asset valuation that includes all transactions costs and, notably, does not take depreciation into account, thereby ensuring the effective replacement of lost assets according to prevailing market values. The importance of establishing practical, yet socially responsible standards for ex ante asset valuation is discussed in greater detail in the section "Land Acquisition Options."

Land Purchases. In some cases wind power developers seek to obtain the needed lands or rights-of-way on a willing seller-willing buyer basis. Not surprisingly, private sector developers tend to be most experienced with this modality of land acquisition.[5] For example, CLP India and Enercon India Ltd., the developers for an 82.4 MW wind power project at Saundatti in the state of Karnataka in India, sought to purchase 135 acres of land for the installation of 103 wind turbines. At Saundatti, about two-thirds of the land was owned by the government, and the remaining one-third was privately owned. Situated on a generally rocky ridge covered with natural scrub vegetation, and not under cultivation, the private land was to be purchased by the developer on a willing seller-willing buyer basis before the project was shelved in 2008.

In a more concrete example, for the AES Kavarna wind farm in northeastern Bulgaria, all land acquisition was conducted on the basis of a voluntary market transaction. The process included identifying and negotiating with multiple landowners across the project site for the acquisition of plots of land that would provide optimal wind farm design and layout. The prices paid for the plots of land exceeded their agricultural land value by a factor of two or more. The farmers will continue to cultivate the land in between wind turbines, extending all the way to the foundations of the wind turbines in most cases, even though part of this land is owned by the project.

The direct purchase approach can also be used to acquire the ROW for transmission lines. The physical displacement of houses and other structures is typically not a major consideration here. For example, the Rio do Fogo project—a 49.3 MW wind power project located in the Municipality of Rio do Fogo in the Brazilian state of Rio Grande do Norte—included construction of a 50 km transmission line. Although the transmission line ran parallel to existing roads and transmission corridors, a new right-of-way was required throughout its length. Enerbrasil, the project developer, engaged a local engineering company to conduct route optimization studies and to negotiate the ROW purchase agreements. The company investigated prices of land, crops, and buildings in the project vicinity and used these as a market reference in order to provide compensation amounts that approximated replacement cost standards in the ROW agreements.

Lease/Rental Payments. Lease or rental payments provide a typical and convenient means for acquiring land for wind projects. The advantage of a leasing arrangement to the landowner is that a direct benefit from the wind farm can be derived while retaining legal ownership of the land. The advantage to the wind farm developer is that it can secure the use of the land for the time it needs to exploit the wind resource while limiting the up-front investment. The agility of lease/rental arrangements tends to make them particularly attractive to private sector developers, though—as is often the case with the spread of any new technology—the lack of broadly agreed upon standards as to what should be included in a leasing contract has made for some ad hoc approaches. One standard that seems to be emerging in different parts of the world involves payment of a fixed price per turbine, plus a one percent profit share, to the owner of each holding affected (S. Krohn, *pers. comm.*). Still, wind power leasing contracts can be quite detailed, with the contracts that have been negotiated for wind farms in Ontario and Quebec, Canada serving as a prime example. In Ontario, farmers whose lands have been identified for wind power development have been advised to enter into agreements that specify everything from the financial terms of the lease (calculation of payments, period of payments, option clauses, and so forth) to the management of local environmental impacts (appearance of wind turbines, procurement of fill material, disposal of gear oil, and so forth) to the cancellation/expiration, exclusivity, and transferability conditions that are part of most legal agreements (OFA 2011). Sample contracts even specify the terms of access by the developer to wind turbines and associated facilities, which is a function traditionally performed by easements.

Easements. Imposition of easements remains the most common way of gaining access, typically for maintenance purposes, to the ROW for transmission lines and access roads connected with wind farms. For example, the Sierra de los Caracoles project in Uruguay employed easements to ensure access to the posts of a transmission line linking the wind farm with the San Carlos transformer station 16 km away. A total of 56 plots of land and nearly as many owners have been affected. Even as the easement was imposed on a plot, compensation was not automatically provided; it had to be requested by the affected landowner via an administrative procedure based on a claim of damages or harm connected to the running of the line through the property. If a claim is lodged, UTE sends its own appraisers to assess the damage and recommend a compensation amount, but the owner has the right to submit his/her own estimate of the damages. A negotiation process then ensues, in which an owner can request an independent appraisal; if that fails to produce a mutually agreed compensation amount, the administrative procedure can be abandoned in favor of a judicial one.

As of this writing, compensation has been requested by six owners in relation to seven plots. In one case, UTE field staff assisted the landowner in the preparation of the compensation request. In another case, a well-off landowner with a larger holding dedicated to cattle ranching hired a lawyer to help him prepare the petition. As with those affected by expropriation in the Caracoles case, most of the landowners subject to easements had access to the information and resources necessary to engage with UTE in arriving at solutions to their displacement that would not put them at a disadvantage. In areas where affected people are not so well-off, however, extra protections have been needed to soften the impact of the restrictions of imposed easements.

Wind Power Development on Indigenous Lands

Indigenous Peoples and Development. The lived realities of indigenous and tribal peoples around the globe are incredibly diverse.[6] Yet a characteristic that many of them share is the tendency to be put at a disadvantage by the forces of purposive development. By now, this tendency has been acknowledged by virtually all development institutions—especially those of a public character—that are active in areas inhabited by indigenous peoples.

Recent Tends in Indigenous Rights. Of the various formal statements with policy implications for the sponsorship of energy schemes (including wind) on indigenous or tribal lands, the UN Declaration on the Rights of Indigenous Peoples is both the most recent and the most comprehensive. A testament to just how far international advocacy efforts led by indigenous peoples themselves have come over the last two decades, the UN Declaration, which was approved by the General Assembly in September 2007, casts indigenous peoples in a role as agents of their own development.[7] It affirms their right to maintain and strengthen their distinct political, legal, economic, social, and cultural institutions while also participating, to the extent that they wish, in "the political, economic, social and cultural life of the State" (UN 2007, Article 5). In situations involving externally directed development projects, it also makes clear that "States shall … cooperate in good faith with the indigenous peoples concerned through their own representative institutions in order to obtain their free and informed consent prior to the approval of any project affecting their lands or territories and other resources, particularly in connection with the development, utilization or exploitation of mineral, water or other resources" (Article 32). Even before the adoption of the UN Declaration, the so-called "FPIC standard" has served as a rallying cry for native peoples and their supporters, who had been growing increasingly frustrated at what they saw as a lack of effective indigenous participation in decision-making concerning native lands and resources.

Indigenous Rights in Practice. The World Bank was one of the first development agencies to have a policy that acknowledges that indigenous peoples, by virtue of their unique circumstances, are exposed to higher levels of risk from development projects,

Photo: Empresas Públicas de Medellín
The wind-swept areas most suitable for wind power development are sometimes the homelands of indigenous people such as the Wayuu community at Jepirachi, Colombia.

and to require measures that both mitigate or manage these risks and enhance benefits-sharing opportunities for them. The most recent incarnation of this policy requires "free, prior, informed consultation," rather than free, prior informed consent. However, taken in combination with other provisions of the policy, particularly the one requiring project sponsors to document that their plans enjoy the broad support of any indigenous communities involved (OP 4.10, para. 11), in practice, there is very little difference between the FPIC standard, as set out in the UN Declaration, and the Bank policy's requirements (World Bank Group 2008). Whatever policy standards are applied, however, increasingly wind power sponsors are finding that merely showing "cultural sensitivity" is insufficient in itself in proposing to develop wind projects on indigenous lands. Rather, the expectation on the part of a number of actors—often not just the affected tribes, but also their advocates in the public and especially in the civil sectors—is that the sponsors will move beyond such clichéd sentiments to the implementation of informed consultation strategies, revenue-sharing arrangements, and social action plans that respond to the needs, interests, and unique histories of the people(s) in question.

Culturally Sensitive Consultations. One of the key elements in successful stakeholder consultations involving indigenous peoples is to earn the trust of their leaders, which can be difficult, especially when the experiences of indigenous communities with outside developers have left a residue of ill-will, or worse. In addition, sometimes it is even a challenge to determine who the legitimate "leaders" are. Often there are intergenerational rifts that pit the young against the old, or differences in the amount of credibility and respect accorded to traditional authorities (such as headmen or shamans) and to formal authorities recognized by the government. A related set of challenges derives from the often fundamental differences in the worldviews, knowledge systems, and communication styles of the indigenous beneficiaries and the more mainstream developers. Moreover, when both sides establish a pattern of interaction that is more about talking **past** or **at** each other than about trying to surpass cultural barriers to communication and reaching a mutual understanding, the long-term viability of any agreements reached is questionable.

The challenges involved in engaging indigenous communities in consultation and participation processes can be seen in the case of the Jepirachi Wind Power project. This project was built on the Guajira Peninsula, a desert area in the northernmost part of Colombia very close to the Venezuelan border. The area is inhabited by the Wayuu, one of the more numerous of Colombia's 80 indigenous peoples.[8] In one community meeting, after the wind farm was built, the Bank-based project team expressed its intention to apply for grant funding for an on-site sustainable electrification project. The grant application process was extensive, and it took longer than expected to finalize a number of details related to the funding. But the result of the community members' expectations having been raised at such an early point was that obtaining access to electricity with the Bank's help became an all-consuming aim for them. As the application process went on, this mismatch in expectations led to some friction between the Wayuu clans (or "*rancherías*") involved and the project team, and to a loss of momentum in the implementation of other aspects of the social action program established for the communities by the project sponsor.

It should be noted that the Jepirachi project otherwise serves as a model example of how consultations with indigenous communities involved in externally induced wind power development could be carried out. The sponsor for the project, Empresas Públicas de Medellín (EPM), took a number of steps in an attempt to garner Wayuu support, including: (i) engaging in extensive up-front discussions, (ii) agreeing to benefits-sharing arrangements targeted at the *rancherías* in the immediate vicinity of the windmills, (iii) supporting social action programs covering all residents in that part of the Wayuu territory, and (iv) undertaking modifications in project design and operational specifications that responded to Wayuu preferences and religious beliefs. The irony and difficulty of external sponsorship of wind projects on indigenous lands is that, even when such significant measures are carried out, there is always the possibility that someone, somewhere will find them wanting. This, in fact, is what happened in the Jepirachi case.

In March 2008, at a regional consultation meeting on climate change mitigation strategies in La Paz, Bolivia, the head of a small Wayuu nongovernmental organization (NGO) called Fuerza de Mujeres Wayuu launched into a sustained and very public criticism of the Jepirachi project. She claimed that consultations on the project had been done with too narrow a group of Wayuu, that the electricity generated by the wind farm was bypassing the communities and going to a mining company located in another part of La Guajira, and that construction of the project was directly linked to a surge in paramilitary activity in the Wayuu territory. A preliminary Bank review of the Jepirachi project's performance up to that time found most of these claims to be exaggerated, and those among the Wayuu who were direct beneficiaries of the project rebutted them forcefully. On several occasions, the Bank tried to enter into a dialogue with the head of the NGO, but these efforts were not met with a consistent response. However, following an intensive effort on the part of the Bank and the sponsor to strengthen the implementation of those parts of the social benefits program that were lagging, the criticisms subsided. While every example of indigenous engagement is different, the lesson for wind power developers is clear: Indigenous communities are not monolithic. Interpersonal rivalries, political rifts, and other forms of instability at the tribal and local government level are often hidden from outside view but on those occasions when they come to the surface, they can cause wind power development agreements to fracture along the same fault lines. In these situations the developers have found that with inputs from qualified, experienced social scientists, they can engage indigenous populations in an appropriately inclusive and participatory way, thereby reducing local acceptance risks to an acceptable level.

For some indigenous communities, wind power constitutes a culturally compatible form of development that can help build sustainable homeland economies. As with the Wayuu in South America, for many tribal peoples in North America, the winds are considered holy, bringing renewal, warmth, and strength. In water-scarce areas, wind is also attractive because it can produce electricity without consuming water. As a result, a number of North American tribes are not just giving wind power serious consideration; they are taking its development into their own hands. For example, in south-central South Dakota (USA), in 2003 the Rosebud Sioux Tribe erected a single turbine on reservation lands to familiarize itself with wind power and the regional electric utility grid system. The Rosebud Sioux Tribe is now planning to install a 30 MW wind farm on its reservation, and several neighboring Sioux tribes are considering following suit (Gray 2008).

Land Tenure Arrangements

Wind power developers have found that the land acquisition and benefits-sharing tasks are much easier in areas where there is a fair degree of clarity over who owns the land needed for the project. Even in cases of localized disputes over exact boundaries, where there are up-to-date land registries and credible means of enforcing property rights (whether the ownership regime is public, private individual, or community), the payment of leases or royalties is fairly straightforward: All one must do is locate the title holder and deliver the compensation or benefit. However, in many parts of the developing world, conflicting and overlapping claims to land are common, and enforcement is frequently lacking, especially in remote areas that are often the richest in terms of wind resources.

Land Tenure Insecurity and Conflict. As is the case with any kind of construction or development involving land taking, where security of land tenure is lacking, a host of conflicts can ensue, putting at risk not only developers' plans, but also the long-term security and economic viability of the communities affected. The risks involved can be seen in the example of a bitter dispute over compensation for land expropriated for two energy projects—a coal-fired power plant and a wind farm in Shanwei, China (see Box 4.1).

Box 4.1: An Example of Land-Related Conflicts in Energy Development in China

In China, farmland has been contracted to farmers since the late 1970s. Many farmers have worked under these contracts for up to 30 years, but under the current constitution, the state can expropriate the land at any time pending adequate compensation.

As China presses forward with rapid development, literally thousands of conflicts have erupted over land seizures, usually because farmers believe they have been inadequately compensated. In 2004, for example, the government claimed that more than 74,000 protests occurred in which at least 100 or more people were involved; most of the protests were over either land disputes or pollution. The problem has been exacerbated by the widening income gaps between China's urban residents and farmers (French 2005). Since 2004, the Ministry of Land and Resources has admitted that in some cities more than 60 percent—and in some places as much as 90 percent—of commercial land acquisitions were illegal (Blanchard 2006).

In some cases, wind power projects have become part of this controversy. In 2002, for example, the Honghai Bay Economic Development Experimental Zone—located in Shanwei near Guandong province, which is close to Hong Kong SAR, China—requisitioned large tracts of arable land, hillsides, and a tidal lagoon for a large-scale development that included construction of a coal-fired power plant and a wind farm. Property seizures for the project led to the displacement of 40,000 residents in Dongzhou village, and subsequently to considerable unrest over the government's compensation and resettlement program (CECC Virtual Academy 2009). In addition, local residents were angry about the pollution that would be caused by the coal-fired plant, as well as plans to fill in a local bay in a way that would adversely affect fishing as part of the wind power development scheme. The long-simmering dispute reached a boiling point in early December 2005 when at least three people were killed and eight injured by riot police seeking to break up a protest at the site (French 2006).

The latest available information indicates that the authorities pressed ahead with the land taking needed for the construction of the electricity-generating projects in Honghai Bay. The coal-fired power plant, for one, was completed and put into operation in 2007, with a capacity of 4.8 million kW and a corresponding investment of 28.6 billion yuan.

IMPACTS ON PHYSICAL CULTURAL RESOURCES

Physical cultural resources (PCR) can be defined as movable or immovable objects, sites, structures, and landscapes and other natural features that have archaeological, paleontological, historical, architectural, religious, aesthetic, or other cultural significance (World Bank OP 4.11). When certain types of development projects, including wind farms, are sited in or near areas where such resources are known to exist, it is quite likely that there will be impacts on them as a result of the construction and/or operation of the projects. At other times, archaeological objects, paleontological remains, or similar examples of such resources turn up unexpectedly as a result of the movement of earth during project construction, making their appearance more a matter of chance. Whatever the case, the project can end up causing damage to objects, structures, or places having value as integral parts of people's cultural identity and practices, as sources of valuable scientific or historical information, and as assets for economic and social development. The assessment of a proposed project's impacts on physical cultural resources is usually an integral part of the EA process, although social assessments and other social diagnostic processes may also be useful for putting the cultural and historical significance of a particular resource in perspective.

Zoning Restrictions and Physical Cultural Resources. In many cases, national or subnational legislation places limitations on what lands can be offered for wind power development. In the Valencia Autonomous Region of Spain, for example, the law specifies that wind farms cannot be located in declared natural spaces protected by the regional government, areas designated for birdlife protection, humid areas classified in the Ramsar Convention, national hunting reservations, forests with significant tree species, or important biological corridors. The law also includes areas with cultural values, and areas classified as having high or protected landscape values, among the no-go areas for wind power development (AVEN 2009). Insofar as the manifestations of cultural values are physically present, developers are finding it necessary to acknowledge the significance that they have for different social groups (usually, but not always, local ones), and to provide for their protection through specific measures included in a project's environmental management plan.

Protecting Physical Cultural Resources in Practice. Drawing freely on UNESCO standards, the legal and regulatory frameworks of most countries, together with the policy statements of multi-lateral agencies such as the World Bank and other project sponsors, provide guidance on the handling and preservation of physical resources having significant cultural or historical value. The application of chance finds procedures and similar measures provided for in these frameworks can help to ensure that wind projects are not developed at the expense of a wind resource area's cultural patrimony. At the Jepirachi project site in Colombia, for example, efforts were made to search for, document, and preserve physical cultural resources during the project preparation stage. Wind turbines and other project installations were located such that they would be at an acceptable distance from cemeteries and other places of cultural significance for the Wayuu. Similarly, in the case of the Uruguay wind farm, the reduction in the number of windmills placed on top of a ridge that served as a natural barrier between large estates also served to limit damage to the foremost example of cultural property existing in the area, a winding stone fence dating back to the seventeenth century.

Photo: Roberto Aiello
This historic stone fence, built during the 17th century, runs adjacent to the recently constructed wind turbine facility in the Sierra de Caracoles of Uruguay.

Social Impacts Particular to Offshore Wind Development

Certain social issues are either less problematic or altogether absent at offshore wind power sites. At the same time, some new issues arise. Socioeconomic impacts arising from offshore wind projects include their visual impact on the coastal landscape and their effect on commercial fishing interests.

Nuisance Impacts. Locating wind turbines offshore has tended to eliminate the noise nuisance. However, aesthetic concerns remain, specifically if offshore wind farms are visible from the shore. Negative visual impacts can be reduced—or even eliminated—by siting wind farms further offshore. However, the costs per kWh produced rise as the distance from the shore is increased. Hence, the social planner is confronted with a trade-off between minimizing negative visual impacts on the one hand and accepting higher costs of power generation on the other (Ladenburg and Dubgaard 2007).

Unless offshore wind farms are located sufficiently far offshore, they generate visual impacts on the coastal landscape. This has the potential to lead to lower levels of acceptance of offshore wind power development. Presently, only a few offshore wind farms are operating worldwide. As a natural consequence, few people have actually experienced these visual impacts on the coastal landscape. Accordingly, up to now, a relatively high level of acceptance of offshore wind farms has been reported in the literature on the subject, although this might not be representative of attitudes in the long run (Ladenburg 2009).

Commercial fishing interests also have frequently raised concerns about offshore wind power projects. In a survey in the town of Horns Rev in Denmark, fishermen were one of the few groups who criticized the existing offshore wind farm. They responded that they would have preferred for the project to have been sited in a different location, to avoid affecting prime fishing spots (Ladenburg et al. 2005).

Fishing interests have also been fairly active critics of the Cape Wind project, off the coast of Massachusetts in the United States. For example, in its comments on the draft environmental impact statement, the New England Fishery Management Council raised concerns about the ability of some types of commercial fishing to continue in the wind farm area. The Council noted that otter-trawl (tow) gear typically extends from about 775 to 1,400 feet, making it very difficult for fishermen to use this gear in between a row of wind turbine generators. Furthermore, a trawler cannot make sharp turns when its net is in the water. To keep the gear from collapsing, it requires a large turning radius, from half a mile to a mile, which is more than the distance between the wind turbines of the project (NEFMC 2008).

Utility of No-fishing Zones. Notwithstanding the often vocal opposition by well-organized fishing interests, the establishment of permanent no-fishing zones in coastal waters has been found often to increase the sustainability of local fisheries, by providing a refuge for fish and other marine life to breed and then repopulate adjacent fishing areas (Salm and Clark 2000). Such no-fishing zones are sometimes deliberately established as marine-protected areas, but they can also result as a by-product of establishing offshore wind farms.

Managing the Social Impacts of Wind Power

Introduction

The second half of this chapter is devoted to good practice strategies for the effective management of the negative socio-cultural and economic impacts of wind projects, with an emphasis on those that are likely to grow in prominence as wind power development is scaled up across the globe. It begins with a discussion of the social criteria to take into account when siting and laying out onshore wind farms and the social analysis tools that can be used in collecting and rendering actionable data on socio-cultural impacts, opportunities and benefits, and risks in relation to a particular wind farm. It continues with a discussion of stakeholder engagement which, among other things, explains the difference between consultation and participation, provides guidance on engaging indigenous peoples, and describes the circumstances under which it is advisable to have a public communication strategy. Strategies for mitigating the impacts associated with the three possible ways of acquiring land for wind projects—expropriation, land purchase, and lease and royalty payments—are then discussed. Finally, the chapter concludes by examining different benefits-sharing options that are relevant to wind projects, before reviewing good practice in the area of socioeconomic monitoring.

Site Selection of Wind Power Facilities

The Importance of Site Selection. As mentioned earlier in this report (see Chapter 2), wind power infrastructure includes wind turbines, access roads, meteorological towers, and transmission lines. Careful site selection of wind power infrastructure is a key consideration in mitigating the social risks and optimizing the social benefits of wind power development.

The Value of a Participatory Approach. In cases where wind power development is externally directed, attitudes toward specific projects emerge from the interplay of the outside agent's actions and the local residents' interests. In this regard, while many people like wind power in general, they do not want wind farms or transmission lines in their "backyards." A participatory approach to wind project siting that takes into account socio-cultural as well as environmental criteria tends to have a positive impact on public attitudes toward a project. As a corollary, such an approach can lead to decreases in public resistance and social conflict over a project. It is particularly important for local residents to be involved in the siting procedure and for project developers to conduct a transparent planning process with high levels of information exchange (Erp 1997).

One consistent finding has been that the public acceptance of wind power tends to increase with prior familiarity, which is itself a function of the level of information exchange. Accordingly, the "not in my back yard" (NIMBY) explanation has been called into question by several studies. For example, a survey in which residents of the United Kingdom were asked about their attitudes towards local wind power plants showed that the longer an individual had been a resident of the community, the higher their level of acceptance was (Krohn & Damborg 1999). The underlying assumption is that the newcomer residents had less information about wind power than the long-time ones, who had been exposed to previous wind power projects near and within their communities.

Managing Visual Impacts. Project developers have many opportunities to study visual impacts prior to construction. Viewshed mapping, photo composition and virtual simulations, and field inventories of views are useful tools for determining the visibility of a proposed project under different conditions, as well as the characteristics of the views. A well-managed project will incorporate a number of techniques into the planning and design process to ensure that project infrastructure is well-screened from view; a case in point is the changes made to the design of the Sierra de Caracoles wind farm in Uruguay to reduce its visual impacts—a process that benefited from the use of photo composition. In general, however, there is no fixed guidance that can help determine if the aesthetic impacts of a project will make it publicly acceptable or not, precisely because of the context-specific variables involved in making such a determination. An emerging issue is the potential for cumulative aesthetic impacts resulting either from several new projects in a particular region or from the expansion of existing projects.

Managing Noise Impacts. Noise from wind turbines is effectively avoided as a problem by ensuring that turbines are located an adequate distance from any human dwellings (other than the wind project offices themselves). A distance of about 300 m from dwellings is usually sufficient to avoid significant noise impacts, although a higher setback—as much as 2 km—might be warranted if the objective is to ensure that nobody would ever hear the turbines from their house. The World Health Organization recommends that outdoor noise sources remain below 55 dB(A) during the day and 45 dB(A) at night. A wind turbine operating below 45 dB, which is at the upper end of the typical range for onshore wind projects, would thus be well below all recognized noise thresholds. At a distance of 200 m, it would produce no more noise than a modern refrigerator. Routinely used industry software exists for calculating theoretical noise impacts on nearby buildings. Standard no. 61400-11, "Wind Turbine Generator SystemsPart 11:

Acoustic noise measurement techniques," of the International Electrotechnical Commission (IEC 2002) provides a method for analyzing the sound output of a wind turbine. Wind power developers should adhere to this standardized method, in order to ensure accuracy in measuring and analyzing the noise levels generated by wind turbine generator systems. A January 2005 revision of the standard also included an analysis on the tone as well as the noise level for each integer wind speed from 6-10 m/s at a hub height of 10 m.

Managing Radar, Telecommunications, and Public Safety Impacts. Careful site selection is the most important tool for avoiding problems with these very site-specific issues. A distance of about 300 m from dwellings is sufficient to minimize any risks to the public from blade or ice throw. Interference with radar and telecommunications systems can also be effectively avoided through careful site selection, including avoiding the installation of turbines within the line-of-sight of the radar or telecommunications facility. Furthermore, wind turbines can sometimes be safely located within the line-of-sight of aviation radar, particularly when additional investments are made to relocate the affected radar, blank out the affected radar area, or use alternative radar systems to cover the affected area. Risks of aircraft collisions with wind turbines and overhead transmission lines are minimized by maintaining adequate distances between airports and wind power facilities. The *EHS Guidelines* provide more details on managing these issues, as well as occupational health and safety.

Wind Farm Layout and Micro-Siting. Just as it is important to consider socio-cultural criteria when determining the larger influence area of a wind project, social criteria are usually pertinent to the definition of the specific siting and layout of the project, the alignment of interconnection roads, and so forth. One obvious example is how the proper placement of wind turbines can avoid problems with shadow flicker. Also, for visual impacts and basic safety reasons, turbines should not be placed in close proximity to dwellings.

Cultural Factors in Siting Wind Farms. At Jepirachi in Colombia, cultural factors were taken into account in determining the location of the wind farm. In its visit to the project, the Study Team noted with satisfaction that EPM, the project developer, had managed to avoid jealous reactions among rival clans in the immediate project area by placing the two rows of wind turbines along either side of the border between the two main *rancherías* in the area, Kasiwolin and Arutkajui. The placement of turbines along each row varied slightly according to a few other criteria, including: (i) the need to maintain a safe distance from cemeteries, considered important stakes in territorial claims by different Wayuu clans; (ii) the possibility that the project might obstruct wind from getting to areas of cultivation, given Wayuu beliefs that the wind plays a role in the fertilization of the land; and (iii) the need to stay away from key archaeological sites, notably traditional Wayuu fireplaces in the area.

Social Planning Tools

Project-Specific Social Assessments. It is still common practice for project-specific environmental impact assessment (EIA) studies to cover social impacts and issues. In this regard, for two of the three LAC case study projects included in this paper (Uruguay Wind Farm and Mexico La Venta II), social issues were analyzed along with the biophysical ones as part of the EIA process. The remaining case study project, Colombia Jepirachi Wind Power, featured a good deal of free-standing social analysis as part of project planning. Increasingly, attention to social issues is being worked into SEAs as well. At the same time, it is good practice to carry out separate social assessments (SAs), particularly for those wind power projects that may have an impact (whether positive or negative) on indigenous peoples or other vulnerable groups. While most countries lack legislation, standards, or capacity to conduct systematic social assessments for projects and programs, a number of international financial institutions have developed good practice standards related to SAs. For example, the International Finance Corporation (IFC) requires social assessment for the private sector projects it finances in emerging markets. This SA is carried out by the project sponsor and, among other considerations, is supposed to provide a complete picture of the expected impacts and associated risks of the project in question.

Entry Points for Social Assessment. Along the same lines, the World Bank's *Social Analysis Sourcebook* (2003) defines SA as a client-directed process that looks at social impacts, opportunities/benefits, and risks in relation to the particular intervention being sponsored. SA establishes the combined analytical and participatory approach that allows the client—or, more typically, consultants hired by the client—to: (i) document ethnic and racial diversity and gender issues that are part of the project setting, (ii) analyze formal and informal institutions in the project setting, (iii) undertake a systematic stakeholder analysis covering both those potentially affected by a project and those who may influence the project's outcomes, (iv) develop a systematic consultation and participation process based on the results of the stakeholder analysis, and (v) identify and address social risks. A social assessment based on these five aspects or "entry points" combines a diagnosis of the socio-cultural context with a deliberate process of stakeholder involvement, and facilitates the incorporation of social sustainability issues into project planning, implementation, and monitoring and evaluation (World Bank Group 2008).

Uses for the Social Baseline. In addition to showing the way to enhance the social benefits and opportunities of a wind project while also helping to reduce its levels of risk, SA can contribute to the project's overall social sustainability in another key way: by establishing a valid, verifiable information base for addressing threats to the reputation of the project and its sponsors. A clear example of this is the La Venta II project in southern Mexico. The December 2003 EA for the project included a social issues section whose demographic data and socio-cultural discussion helped to establish the Zapotec indigenous heritage of the residents of Ejido La Venta. To some extent, the recognition of this heritage influenced the communications, consultation, and benefits-sharing actions of the project sponsor, CFE. In addition, it also provided CFE the means with which to rebut certain claims made by the NGO Intermón Oxfam in an investigative report that was critical of La Venta II. For example, one of the report's assertions was that project officials disregarded the collective decision-making process that is a typical feature of an indigenous *ejido*, opting instead to negotiate contracts with the *ejidatarios* on an individual basis (2009). For its part, CFE was able to show that, before there was any discussion of contractual arrangements, it had received numerous statements in support of the project through traditional community assemblies. This helped to bolster arguments for the project's social legitimacy while providing reputational protection not only for CFE, but also for the Bank's Spanish Carbon Fund, which provided carbon finance credits for La Venta II.

Because developing wind power on indigenous lands can be sensitive, careful planning is needed. Making sure that the appropriate type and amount of social analysis is conducted is only the beginning. When the World Bank supports a proposed project affecting or involving indigenous peoples, it requires the project sponsor to prepare and implement an Indigenous Peoples Plan (IPP). Such a plan is designed to set out the measures by which the project sponsor will ensure that: (i) any indigenous peoples involved receive culturally relevant social and economic benefits; and (ii) any adverse impacts on said peoples on account of the project are avoided, minimized, mitigated, or compensated (OP 4.10, para 12). The IPP serves as an appropriate vehicle for summarizing the findings of the social analysis conducted, including the outcomes of the free, prior, and informed consultation process. This, in turn, provides a critical opportunity for documenting the nature and extent of community-level support for the proposed project. The IPP should also establish a framework for continued consultations with the communities concerned throughout the construction, operation, and maintenance phases of the project. These are some of the topics covered in the following section of the paper.

Stakeholder Consultation, Participation, and Communication

Transparency on Trade-Offs. A critical part of public consultation in the context of energy development is educating the public and decision-makers about the full range of trade-offs, impacts, and benefits associated with different technologies, including wind. From the earliest phases of a project, information on the implications of proposed wind power development for both the natural environment and local people should be adequately disseminated, so that all trade-offs can be properly considered—as through a public hearing—in the course of the decision-making process. Typically, this is the objective of policies and procedures on the disclosure and dissemination of project-related information, which find their most direct expression in EA legislation in different countries. While such information-sharing is a necessary prior step, international good practice on stakeholder consultation in projects indicates that it is, in itself, not sufficient.

Promoting Full Participation. For all relevant stakeholders to be as fully invested in a proposed wind project as possible, the consultation process cannot be one way, as when the project developer provides project information to likely beneficiary groups or adversely affected people without any expectation of a response. Nor should it be limited to two-way interactions, as when the developer shares information on a possible wind farm layout in order to solicit the inputs of area residents. Often the most socially equitable and sustainable outcomes are achieved when public participation is sought. "Participation" denotes a more meaningful and informed process of involvement with stakeholders, in which key groups actively participate in defining and implementing the project or cluster of projects in question (World Bank Group 2008). Where the private sector is involved, participation may be seen as one of the more advanced steps in a process of engagement with stakeholders that could culminate in the formation of strategic partnerships around the proposed project or venture (IFC 2007). Public participation tends to be most effective when it is facilitated in a structured and sequenced manner, as outlined in Box 4.2.

The benefits of a "consultation-plus" approach to stakeholder engagement can readily be seen in recent wind project experience. In the case of Jepirachi in Colombia's Guajira Peninsula, the project sponsor, EPM started by conducting exhaustive upstream consultations on the project design and the proposed social benefits plan. This was confirmed in interviews with EPM technical staff and with members of affected Wayuu communities during a visit of the Study Team to the project site. The consultations were structured in such as way as to encourage the participation of a range of affected

---

Box 4.2: Steps to Effective Stakeholder Participation

---

1. **Stakeholder analysis.** Define relevant groups and actors, including not only those affected positively or negatively by a project, but also those groups such as NGOs, politicians, and others who are in a position to influence project outcomes.

2. **Prior meaningful information and disclosure.** Provide relevant information about the proposed Stakeholder groups taking account of their interests, needs, and likely concerns. As required by the policies, disseminate draft project documents and plans in locations and languages accessible to key stakeholder groups.

3. **Transparency in process.** Explain and clarify to stakeholder groups the decision-making process and how stakeholder views will be incorporated.

4. **Appropriate forums and methods.** Organize the consultation and participation process in ways that capture the views and concerns of different groups, in manners and forums appropriate to them (such as a combination of focus groups, workshops, interviews with key informants, separate events for men and women where needed, and so forth)

5. **Documentation.** Document dates and events, participants, and the views, recommendations, and concerns expressed by different groups.

6. **Decisions.** Document decisions taken on the basis of stakeholder inputs, and why they are justified.

7. **Feedback and future involvement.** Provide timely feedback to participants with a summary of decisions taken. Where appropriate, provide opportunities for continued engagement and dialogue.

8. **Redress and appeal**, with sufficient independence and authority to mediate or provide support to affected populations as appropriate (for example, Ombudsman function, other institutions).

Source: Reprinted from World Bank Group 2008.

---

indigenous stakeholders, and to earn EPM the trust of community leaders—no small feat, given the Wayuu's fresh memories of negative experiences with *arijuna* ("whites" or "outsiders" in Wayunaiki) such as the former managers of the nearby Cerrejón coal mine. The consultation process resulted in the development of two separate community benefits programs, which were designed according to sensible principles, and within time frames that respected Wayuu decision-making processes.

Seeking Expert Inputs. A key element in building trust among indigenous stakeholders and helping to ensure that wind projects sited on indigenous lands are based on relevant and accurate information is to employ socio-cultural experts who are intimately familiar with the project setting and the social groups in question. Once again, Jepirachi serves as a good practice example in this regard. Not only did EPM have an experienced anthropologist on board to help guide the initial consultation process, but the company also reorganized the staffing on the project so that this person and the environmental specialist could remain involved in community engagement issues after the wind farm was built, even though they formally belong to EPM's project planning division. EPM also rounded out its expertise in the social ambit through the hiring of experts like Wilder Guerra, an anthropologist who is half Wayuu. Guerra, the prizewinning author of numerous pieces on Wayuu history and culture, supported the project by preparing an intercultural relations manual.

Dealing with Sophisticated Stakeholders. Similar principles apply in settings with a lesser degree of socio-cultural diversity than what was observed in projects like Jepirachi. Where stakeholders are sophisticated or where there is the potential for a lot of conflict, it is advisable to have a targeted public communication strategy. In this regard, a lack of communication between residents in a wind resource area on the one hand and developers, equipment suppliers, local bureaucrats, and elected representatives on the other, may become a catalyst for social mobilization and protest actions against specific projects (Damborg 1998). Conversely, information-sharing and dialogue from the very beginning of project planning is crucial for achieving social acceptance. Often people want assurances not only that they are being heard, but also that their views and preferences are being respected. If they accept a consultation and participation process as legitimate, they are much more likely to go along with final decisions taken even if they do not agree with all of them (World Bank Group 2008). Finally, socioeconomic or political pressures, local policy agendas, and the extent and form of organization of the local population and opinion leaders are likely to provide the best indication of how wind power can be introduced so that broad support for project developers' plans, as well as a measure of social equity, can be achieved.

Land Acquisition Options

Expropriation or Purchase. As noted earlier, the land acquisition needs of a typical wind farm make it highly unlikely that physical displacement and resettlement would be necessary. It is much more common for wind turbine facilities and ancillary infrastructure to affect a relatively small portion of an owner or occupant's overall plot, giving rise to economic displacement. When such displacement is not covered by lease/rental arrangements or payments connected with easements, the land has to be acquired either through a formal expropriation process, or through outright purchase. Several factors typically influence a developer's decision regarding which of these two land acquisition options is likely to be more effective.

Applying Land Expropriation. The first factors to consider are related to the overall context for the wind power project, starting with its enabling environment. Not surprisingly, where a host country features an adequate and enforceable legal, regulatory, and policy frameworks for land acquisition, exercising the state's power of eminent domain might be the most appropriate vehicle for land acquisition. On the other hand, where active, well-functioning markets for land exist and the affected people use such markets, implementing a direct purchase modality might be more appropriate; this is usually more socially desirable as well, as it removes the compulsory element from the equation. An interesting aspect of most wind farm layouts is that they can accommodate irregular rows of turbines, which sometimes happens when there are hold-outs among a group of landowners. Such was the case at La Venta II in southern Mexico: The decision of a handful of disaffected *ejidatarios* not to participate in the project created "holes" in a couple of the rows, but this has had a negligible effect on the overall operation of the wind farm. A win-win situation ensues in which both sides get what they want: development of the project can proceed, but the owner of a plot in the middle of it can elect to opt out. The possibility that social conflicts will arise later as a result of "envious" actions by the owners who opt out—as was the case in La Venta II—can be avoided through payments of some share of the royalties generated by the operation of the wind farm to all those having holdings within the wind farm area, whether they choose to have turbines erected on their plots or not (S. Krohn, *pers. comm.*).

Additional factors specific to the local socioeconomic and cultural context should also be taken into consideration. It may be important to consider, among other things, the prevailing tenure regime in the project influence area (indigenous peoples' communal lands cannot be purchased or expropriated); the socioeconomic profiles of the landowners or occupants affected (acquiring the productive lands of farmers through direct purchase may lead to impoverishment if they lack the experience or training needed to parlay those sums into new income-generating activities); and the history of relations between local populations and outside developers. Wind power developers are therefore advised to make use of the inputs of social scientists, community relations specialists, and others with similar expertise—not to mention of area residents themselves—during the earliest stages of a proposed project, in order to decide on a practical, fair course of action.

Pursuing Land Purchases. As noted earlier, the nature of the developer itself is also likely to influence the decision. Private companies tend to be more comfortable with facilitating willing seller-willing buyer transactions, whereas public utilities experienced in the activation of the formal expropriation process often choose to go that route. In some cases, however, it may be in a utility's interest to acquire the land needed for a wind energy project by purchasing it, as when local law allows the costs involved to be included in the utility's rate base (the investments and expenses the utility is allowed to recover from customers) (Winrock International et al. 2003).

In the unlikely event that construction of wind turbine facilities and/or ancillary infrastructure should require involuntary resettlement, it is important for the developer to proceed with care. The Bank's experience is that physical displacement and resettlement can result in "long-term hardship, impoverishment, and environmental damage unless appropriate measures are carefully planned and carried out" (OP 4.12 para. 2). Among the measures that the Bank requires in the projects that it finances include, among others, carrying out socioeconomic surveys and conducting censuses, developing an entitlement framework, defining a consultation and participation strategy for affected persons, and preparing, implementing, and monitoring mitigation plans centered on the replacement of lost assets and the restoration of incomes. Further information on the full range of recommended measures is available in the World Bank's *Involuntary Resettlement Sourcebook* (2004).

Compensation Standards. The payment of compensation (*indemnización* in Spanish) is the most common means of asset replacement in situations where a limited amount of land needs to be expropriated for a project. While asset valuation methodologies are a well-established feature of most property law regimes, wind power developers operating in countries where compensation amounts offered are based on the cadastral values of the affected lands should think twice before proceeding on this basis. By contrast, compensating people at "replacement cost" tends to achieve more satisfactory results.[9] In addition to depending on clear and consistent methods for the valuation of affected assets, the calculation of replacement cost ensures that the assets are not downgraded in value (as by taking depreciation into account), thereby allowing for their effective replacement according to prevailing market values. It also ensures that all transaction costs are included. This sets a higher bar for compensation than what is legally provided for in many developed countries. In the World Bank's experience, valuing assets according to a replacement cost or equivalent standard goes a long way toward promoting the social sustainability of a development project, especially in cases of: (i) economic displacement affecting poor populations, even if they are not required to move, or (ii) physical displacement and resettlement. But regardless of which approach to asset valuation is ultimately chosen, the criteria involved should be applied consistently to avoid generating conflicts among neighbors.

Imposing Easements. Met towers and above-ground transmission lines connecting a large wind farm via substations to the distribution grid do not require land acquisition *per se*, except for the tower bases. Acquisition of the ROW for the transmission lines may be done via direct land purchase, but more often it is done via the imposition of easements. This normally involves the payment of an easement fee as compensation for restrictions on the residential and commercial use of the land under or near the lines, and for periodic access to the towers by the maintaining utility. Additional payment for crop damage might also be added (World Bank Group 2004). Most types of previous economic activities can continue. In most cases, no compensation is paid for decreases in property values as a result of the tending of transmission lines, although it is good practice to seek out ways of addressing the economic, safety-related, or aesthetic concerns of landowners whose properties are traversed by the lines. For example, landowners affected by easements needed for the transmission line linked to the Caracoles wind farm in Uruguay have been regularly and conscientiously briefed by UTE field staff on their rights and options under the law. Among these are the right to request that their lands be formally expropriated, if the owners believe that the presence of the lines have devalued their holdings to the extent that they are no longer economically feasible.

Compensating Temporary Impacts. It is also good practice for a developer to make prompt and adequate payments for "related damages," which can be defined as unforeseen or additional damages tied to temporary impacts during the construction of a wind project. An example of this would be accidental property damage resulting from the construction or widening of an access road for transporting turbine towers to a wind farm site. The need to make such payments is usually also provided for in the relevant legislation of many countries, if not in the project developer's operating policies.

Benefits-sharing Arrangements

Compensating people who have lost land and other assets, either temporarily or permanently, on account of land acquisition for wind power development is an important step. However, project developers also do well to concern themselves with delivering tangible benefits to area residents, with the aim of not only securing the lands needed for a project, but also sustaining local livelihoods, improving incomes, and enhancing local people's skills and knowledge. By going beyond those who are immediately and adversely affected by the construction and operation of a wind farm, sponsoring companies can find themselves in a much better position to manage public perceptions, improve overall community relations, and engage in effective risk management. Where public utilities and other government entities are involved, taking an approach to scaling up wind power that involves the broad sharing of benefits can serve as a means of supporting regional economic development, achieving equity in the distribution of public resources, and fostering political support through enfranchisement.

Benefits-sharing Options. Not only should benefits be conceived of as standing apart from compensation for adverse impacts; they can and should go beyond the payments that are usually at the center of the land-acquisition strategy for a wind farm. A view of potential benefits that focuses narrowly on financial streams misses other potential benefits that can be of significant value to local stakeholders. In looking at possibilities for benefits-sharing, the idea of "stakeholders" itself needs to be considered more broadly, in recognition of the ties that often exist between and among social groups in a wind farm's larger influence area and of the strategic importance of engaging those actors who have the power to affect a proposed development, not just those who are affected by it. Agreements on sharing financial benefits with concerned stakeholders, as through the payment of royalties, address the distribution of economic rent from the operation of a wind farm. Joint decision-making on the location, design, and construction of wind power facilities can offer additional benefits such as the clarification of property rights for host communities during project preparation or employment opportunities. Other ways to maximize the benefits of a wind power project can include the design and implementation of local benefits programs and the promotion of community ownership of wind farms, which are discussed below.

Clarifying property rights, where required, serves as a particularly powerful example of a win-win approach to social issues management in wind power development, especially when it is treated as a precondition to such development. Local beneficiaries can feel validated and supported in knowing that what may be a long-standing concern of theirs is finally being addressed, while the developers of a given project can gain reassurance from knowing that it will not be delayed or even undone by competing or overlapping claims to the lands needed.[10] One case that readily illustrates the advantages of taking this proactive approach to conflict resolution is that of La Venta III, located in the town of Santo Domingo Ingenio in the Mexican State of Oaxaca. Setting aside the land needed for this wind farm, which will have an installed capacity of a little over 100 MW once it is complete in late 2010, has been facilitated by the clarification of property rights within the existing land tenure regime, which is based on collective ownership under the *ejido*. The idea is to preempt the types of disagreements over the delivery of lease payments that have emerged in La Venta II by working with the relevant state actors to give the former *ejidatarios* in Santo Domingo individual certificates to the plots they have been cultivating, which is precisely where the wind turbines would be installed. While the national electricity company CFE is not the developer in this case, its staff is nonetheless providing support to this process on the strength of their experiences with land-related conflicts and *ejidatario* relations.

LEASE/RENT OR ROYALTY PAYMENTS

Types of Payment Arrangements. A common way of sharing benefits in connection with wind power development is to provide lease/rent or royalty payments to the owners of the land needed for a wind farm's operation, above and beyond what they would receive as compensation. The most common type of contract for doing so provides for the payment of royalties with a percent of gross revenue (often 1 percent), or a percent over billing. This arrangement provides an incentive to both the developer and the landowners to maximize the productivity of the wind farm. It is also easy to verify. To prevent the landowners from receiving lower-than-expected payments due to factors such as the technical failure of the turbines, royalty schemes usually also include a guaranteed minimum payment.

Other types of payment structures include lump sum payments—a relatively uncommon arrangement since it preempts the possibility of an ongoing economic agreement between the developer and landowner—and a flat or fixed fee arrangement. Calculated on the basis of individual hectare or wind turbine, the latter arrangement provides both landowner and developer with some certainty regarding future income or payment streams. Among its disadvantages are that the payments do not reflect the actual revenue generated, and the incentive for the landowner to cooperate with the developer to ensure optimum power generation is eliminated. For a summary of the advantages and disadvantages of these different types of payment schemes, see Table 4.2.

Leasing Contract Provisions. In cases involving multiple landowners, developers usually take one of two approaches: (i) basing payments on the power generated by specific turbines located on the individual plots of land, or (ii) basing payments on the average output of all the turbines in the project, multiplied by the number of turbines located on an individual plot of land. The second option is easier to verify and document

Table 4.2: Comparison of Different Revenue-sharing Arrangements

Arrangement	Advantages	Disadvantages
Lump Sum Payment	• Landowner: Source of immediate cash • Developer: Does not have to provide payments in subsequent years	• Landowner: Does not provide steady income stream • Developer: Must provide lump sum up front

Flat or Fixed Fee (per turbine/ hectare)	• Landowner:
	  – Provides steady, predictable income stream
	  – Protected in years of low electricity and/ or revenue
	• Developer: Does well in high-production/ revenue years
	• General:
	  – Can be used to compensate a landowner for use of land for an access road crossing the property, even if there is no turbine installed on the land.
	  – Clarity and transparency: Easy to verify
		• Landowner: Forgoes potentially higher, if fluctuating, level of income associated with royalty payments
		• Developer: Expenses are harder to bear in years of low electricity and/or revenue
		• General:
		  – Payments do not mirror actual revenue generated
		  – Eliminates the economic incentive for the landowner to cooperate with the developer to ensure optimum power generation

Royalties	• General:
	  – Take into account varying productivity
	  – Give landowner incentive to work with developer to place the turbines on the most productive locations
	  – Give landowners and developers incentives to ensure continuous power generation
	  – Easy to verify if based on gross revenue
		• Landowner: Difficult to verify electricity and revenue generated by each turbine:*
		  – Individual turbine generation information is hard to get
		  – Individual monitors on turbines do not reflect the energy sold because they do not account for energy losses in the electrical system
		  – Developers generally do not like to share turbine productivity data

Royalty/Minimum Guarantee combination	• Same as above, with additional benefits from an up-front fee or a minimum guarantee	• Same as above

Source: Reprinted from Winrock International et al. 2003.
Note: * In the United States, information about the amount of power generated by a facility is publicly available from grid operating managers or the utility purchasing the power. Even so, such information does not indicate how much is generated by each wind turbine.

and carries the least risk for the landowner. Other provisions in leasing contracts are designed to ensure that there are no misunderstandings during the life of the project. For example, activities such as ranching or farming that can be conducted simultaneously on the land around the turbines need to be specified (Winrock International et al. 2003).

Range of Payments. In the United States, the range of payments under the royalty scheme is typically between 1 and 4 percent of gross revenue, with most in the 2-3 percent range. In Latin America, most payments also were in the 2-3 percent range. For both flat-fee and royalty payments, the average annual payment was US$2,200 per MW, with a range of US$1,200 to US$3,800 per MW. It should be noted that these payments vary depending on the context. Projects in regions with superior wind power resources and the potential for a higher density of turbines—such as the southern Tehuantepec Isthmus region in Mexico, the site of the La Venta II Project—may produce higher revenues per hectare and higher payments to landowners (Winrock International et al. 2003). This provides an important incentive to these landowners—not to mention to other local groups that stand to gain something from employment opportunities, local benefits programs, and similar spillover effects—to become active proponents of wind power development in their regions.

LOCAL BENEFITS PROGRAMS

The Importance of Local Benefits Programs. In addition to agreeing to share financial benefits, some wind power developers take the extra step of engaging local communities and other relevant stakeholders in the design and implementation of collective benefits programs. Such programs, which are often based on the provision of services, construction of public infrastructure, in-kind donations, and the like, tend to be essential elements of corporate social responsibility and/or community relations strategies pursued by private sector developers. Furthermore, pursuing local benefits programs can sometimes lead to unanticipated goods, as when developer assistance to communities in the articulation of their needs leads to the restoration of government presence in a region, and a concomitant increase in the provision of public services. In areas where there are multiple developers working at the same time, they should explore ways of coordinating their efforts on community investment in order to enhance local benefits.

Local Benefits and Carbon Finance Projects. Public or quasi-public sector actors interested in wind power projects might also be interested in enhancing their local credentials through the sponsorship of local development programs, but other incentives come into the picture here. Many of these entities—including those at the center of all three cases discussed in this study—are relying on carbon finance credits to help them overcome financial, regulatory, and other barriers to wind power development. The demonstration of sustainable development impacts—which in practice has often meant providing for local benefits for the larger host community, not just those people who are immediately affected—is a requirement for all projects registered with the Clean Development Mechanism (CDM). One World Bank–administered carbon fund, the Community Development Carbon Fund (CDCF), even goes so far as to provide for payment of a premium on the value of emission reductions, for use in funding the implementation of local development activities.

The willingness of EPM, the sponsor of the Jepirachi project in Colombia, to consider the development interests and goals of the Wayuu indigenous communities in the project's area of influence resulted in a long-term collaborative relationship with those same communities, to the benefit of all involved. The public corporation explicitly rejected the idea of negotiating a one-off set of benefits as "compensation" for the extended use of community lands for the wind farm. Instead, it engaged the local Wayuu in the development and application of an Institutional and Community Strengthening Plan (*Plan de Fortalezimiento Institucional y Comunitario*), resulting in a more continuous delivery of collective benefits for them than a one-time negotiation would have. Under the original arrangements, the participating Wayuu received easement payments and compensatory benefits in a first phase. The social benefits plan for this phase (*Plan de Gestión Social*), dating to September 2002, was very complete and well considered. The second phase involved the implementation of a full-fledged social action program (in Spanish, *Programa Social Adicional*). The activities of both phases were designed to become "self-managed" by local communities and institutions within a medium-term timeframe, which has happened to a great extent.

A good example of corporate social responsibility activities involving employment as a central benefit can be seen in the case of two IFC–Netherlands Carbon Facility–sponsored wind farms in India, where the private sponsor is Enercon India Limited (EIL). EIL has built new access roads and rebuilt local schools, clinics, and temples in the vicinity of its wind farms and manufacturing plants. In particular, the company played a catalytic role in the development of the Daman light industrial area in the mid-1990s, which now provides employment to hundreds of local villagers. At both the Rajasthan and Karnataka wind farms, EIL has generated employment opportunities for local villagers, to provide security and other support services to these installations. The company has also expressed an interest in participating in the *IFC Against AIDS* program. This program includes awareness training, community engagement actions, and a Corporate Citizenship Facility that provides financial support to clients like EIL of up to 50 percent of eligible AIDS prevention–related projects.

LOCAL EMPLOYMENT OPPORTUNITIES

Employment During Construction. Experience has shown that the greatest benefits in terms of job creation from scaling up wind power have occurred during the construction phase of a project. At the Tejona wind farm in Costa Rica, for example, 75 percent of the people employed during construction came from the district where the windmills are sited, and an additional 16 percent came from the surrounding province. During the period of maximum activity, about 200 people worked on the project. Local training programs related to agriculture and environmental quality issues such as waste disposal have been implemented to help generate more local jobs.

Employment During Operation. Once wind farms become operational, local employment opportunities tend to be much scarcer. Many skilled workers, like the turbines themselves, are likely to come from foreign countries. An example of this can be seen in the case of the Uruguay Wind Farm project, where technicians from the Spanish firm Eduinter and the Danish firm Vestas are responsible for the project's operation and maintenance, respectively. One way to improve job opportunities under these circumstances would be to give local workers the chance to receive on-the-job instruction in a variety of maintenance tasks from foreign technicians who have gone through one of the training programs offered by the major wind turbine manufacturers.

Offshore Wind Farms. The employment benefits of offshore wind projects were recently calculated using input-output model data. Using the Horns Rev Offshore Wind Farm as a model, the calculations show that the establishment of an offshore wind farm with 80 two MW turbines creates a total of around 2,000 person years of domestic employment over the construction period. A tentative estimate indicates that up to one quarter of this job creation will be at the local level. Operation and maintenance over the 20-year lifespan of the facility will create an additional 1,700 person years of employment—or 85 person years on average on an annual basis. It is expected that three quarters of this will be at the local level (Ladenburg et al. 2005).

COMMUNITY-BASED WIND POWER SCHEMES

Rapid Spread of Community-Based Wind Power Projects. This study has focused overwhelmingly on the main environmental and social sustainability issues that go hand-in-hand with the growth of wind power, where such growth consists of the push by private companies and public utilities to build large-scale, grid-connected wind farms. However, no discussion of scaling up wind power would be complete without a consideration of the rapid spread of community-based wind power projects, especially in the United States and Europe. Although community wind projects can be of any size, they are usually commercial in scale, with capacities greater than 500 kW, and are connected on either side of the meter. Community wind includes both on-site wind turbines used to offset the owner's electricity loads and wholesale wind generation sold to a third party. In the former case, community-level wind power development is effectively taking the typical wind farm benefits-sharing scheme one step further: under this type of scheme, the main benefit that accrues to the owner of the wind project is the electricity generated by the project itself. In this regard, social issues relating to community consultation, communication, land acquisition and the like cease to have the same prominence since the project developer/owner and the local beneficiary are one and the same.

Other Advantages of Community Wind Power. Local communities in the United States have started to undertake wind development as a way to diversify and revitalize rural economies. Schools, universities, farmers, Native American tribes, small businesses, rural electric cooperatives, municipal utilities, and religious centers have all installed their own wind projects. Community wind is likely to continue advancing wind power market growth because it offers the following advantages:

- **Minimization of impacts:** When developed on a small scale, community wind projects impact ever-smaller land areas and often do not require ancillary infrastructure such as transmission lines, as they can connect directly to the existing local distribution grid. Also, everything else being equal, one or two windmills have a much less chance of causing harm to birds and bats than a large array of them.

- **Strengthening of communities:** Locally owned and controlled wind power generation not only serves as a welcome source of new income for farmers and landowners, but can also substantially broaden local tax bases in jurisdictions where there is a mechanism for local income taxation, thereby benefiting entire communities.
- **Mobilization of support:** Local ownership broadens support for wind energy, engages rural and economic development interests, and builds a larger constituency with a direct stake in the industry's success. Local investments with positive local impacts produce local advocates.

Wind Cooperatives in Europe. Local ownership of wind power projects—in the form of cooperatives—has played an important role in the development of wind energy in Scandinavia and elsewhere in northern Europe. For example, wind cooperatives and individually owned wind turbines have dominated the wind power sector in Denmark, with shares up to 85 percent of installed capacity for the past 30 years. Cooperatives have also taken hold in Germany and Sweden, although later and without the dominant position in the market that they have in Denmark. In Germany, individual investors have typically held fairly large shares, whereas Danish and Swedish cooperatives have mostly had very small shares. As of June 2009, the typical investor in Denmark had about 3,300 euros (US$4,700) invested. Due to the highly localized ownership and the small investment per individual, the Danish model is particularly relevant to the situation in many developing countries.

Danish wind cooperatives are normally organized as unlimited partnerships, where each investor owns a fully paid-up share of a particular wind turbine or a group of wind turbines. Under this form of organization, the cooperative itself has no debt, while the individual may finance part of his/her investment out of her/his savings or by using his/her shares as collateral for a bank loan. In the absence of indebtedness, the unlimited partnership in practice carries no default risk for the individual investor (other than losing her/his own share). The success of the Danish wind cooperatives in obtaining local support for wind development has been quite remarkable: nowadays wind turbines are highly visible across the country, and some 25 percent of Danish electricity consumption today is covered by wind energy. Further details on how this particular modality of wind power development has functioned in Denmark are provided in Box 4.3.

Leveraging Public Financing for Social Sustainability

Diversifying Incentives. Apart from the CDCF premium on emissions reductions that allows for the integration of local benefits programs into wind projects supported by this particular carbon fund, there are few examples of fiscal, policy, and similar incentives for helping ensure that social opportunities and values are enhanced in connection with wind power development. In fact, it is more common to see examples where wind-related fiscal incentives with a social bent have been worked into the legal and regulatory framework for renewable energy development. A description of how such an approach has facilitated local ownership of small-scale wind power projects in Minnesota can be found in Box 4.4.

Box 4.3: Wind Power Cooperatives in Denmark: An Applicable Model for Developing Countries?

Initial support for wind power cooperatives in Denmark was largely based on the need to secure local support, at the municipal level, for obtaining planning permission to install wind turbines. Denmark has had policy targets for increasing the share of wind power in the electricity mix for more than 30 years, and an important way of obtaining local acceptance of wind turbines has been to support local ownership. In fact, many sectors—agriculture, savings banks, credit unions, and the power sector—were originally organized as cooperatives facilitated this form of organization in the country. In addition, local ownership of wind turbines has found support in the existing legislative and institutional setup.

Public policy support has primarily consisted of simplifying tax regulations for small investors in wind turbines by allowing them to be taxed on only 60 percent of their gross revenue from sales of electricity. The first 400 euros (US$570) of gross revenue per shareholder is tax free, which is an attractive tax break for shareholders in wind turbines.

Among others, there are several driving forces for investors in wind cooperatives, including the following:

- **Wind cooperatives provide good economic returns combined with relatively low risk due to small shares.** Yields must generally be competitive with returns on financial investments in order to provide adequate incentives.
- **Investors perceive that they personally contribute to environmental improvements through their actions.** For instance, in Denmark, wind turbine investors perceive that they contribute to a cleaner form of electricity production. The sense of individual responsibility toward the environment is particularly strong in many northern European countries, including Denmark.
- **People invest in local production and generate local tax revenues.** For almost 20 years, Danish legislation limited ownership in wind farm cooperatives to people in the municipality in which the turbine was located (later including neighboring municipalities). Local residents had to sell their shares when leaving their municipality. When the legal requirement of only local ownership was dropped, a substantial number of new investors from urban areas could invest, but their investment was often resented by local populations, in particular members of established wind cooperatives. This tension between local and non-local investors still exists.
- **Investors tend to view wind projects favorably.** Whether the arrival of a wind power project is seen as being a potential boon or a burden by the local population is partly determined by the applicable ownership arrangements. For example, opinion polls have shown that shareholders in wind turbines systematically have a more positive attitude toward the appearance of wind turbines in the landscape than non-owners.

Can the cooperative model work in developing countries? For example, the Mexican *ejido* system, on the surface, appears to be amenable to cooperative wind ownership schemes. However, this form of cooperative governance and its interaction with municipal regulatory roles are quite complex. In addition, currently Mexican institutions are not in a position to support such arrangements. Therefore, for the cooperative model to work effectively in developing countries, a concerted effort would have to be made to adapt European cooperative ownership models to developing country circumstances—unless it is possible that home-grown local ownership models could emerge on their own, or otherwise be cultivated.

Box 4.4: Creating Incentives for Community Wind Development in Minnesota, USA

Starting in the early 1990s, Minnesota took major steps to encourage the development of renewable energy by requiring the state's largest utility, Xcel Energy, to integrate wind power into its operations. The target was 425 MW in 1994, 825 MW by 1999, and 1,125 MW by 2003. The result was the creation of a reliable wind power market in the state which, in turn, helped wind power find its way into many areas of Minnesota's economy, including construction, operation, maintenance, and engineering. It also forged the path for development of enabling rules that other states and counties use as models for writing regulations.

Community wind power began in Minnesota in 1997, when local wind advocates worked with the legislature to create the Minnesota Renewable Energy Production Incentive (REPI). Local ownership was a priority for those who created this incentive, which paid US$0.01 to US$0.015 per kWh for the first 10 years of production for projects smaller than 2 MW. In the beginning, small-scale wind developers had to individually negotiate with utilities for interconnection and PPAs. It was not until a special community wind tariff was created in 2001 as part of Xcel Energy's merger settlement, that community wind projects really became feasible. This tariff established a set power purchase rate of US$0.033/kWh and standard procedures for interconnection for wind projects below 2 MW.

Following the creation of the special tariff, the initial Minnesota REPI allocation became subscribed quickly. A second round was fully subscribed within six months. The pairing of these complementary policies allowed the community wind market in the U.S. to grow and thrive.

Source: DOE 2008.

Socioeconomic Monitoring

The Function of Socioeconomic Monitoring. The way in which wind farms are operated tends to provide the most important opportunities for sound environmental and social management, and solid monitoring arrangements are a key part of any wind farm operating plan. Just as it is important to include a bird and bat monitoring protocol in the EMP for any wind power project where bird and bat mortality has been identified as a possible issue, socioeconomic monitoring is also necessary to ensure that compensation for temporary or permanent land acquisition is paid, that royalty sharing agreements are being complied with, and that goods and/or services promised under a local benefits program are being delivered. Furthermore, in the unlikely event that the acquisition of land for a project involves physical displacement and/or a significant adverse effect on assets, it is good practice to conduct a post-displacement evaluation, to confirm that those assets have been successfully replaced, and that any impacts on livelihoods (economic displacement) have been effectively addressed.

Where carbon finance is involved, socioeconomic monitoring indicators are occasionally included in the Monitoring and Verification Protocol (MVP), which is a standard part of documentation required for the registration of a carbon finance operation with the CDM. This would place at least some of the responsibility for monitoring these indicators with UN-accredited Designated Operational Entities, whose main job is to validate proposed CDM project activity and certify greenhouse gas (GHG) emissions reductions. This is not established practice, however; moreover, wind power projects that involve the application of World Bank safeguard policies or IFC performance standards typically require attention to a wider range of socioeconomic and cultural variables than what is included in the typical MVP. In these cases, the social monitoring arrangements for a wind power project should be discussed and designed as part of the SA process during project preparation, and formally provided for during the post-construction period in the Local Benefits Plan, Indigenous Peoples Plan, or similar.

An example of where the monitoring requirements of carbon finance regulators and of the World Bank have been satisfied in a mutually reinforcing way can be seen in the Jepirachi Wind Power project in Colombia. The creation and maintenance of a monitoring system, establishment of viable "sustainable development" indicators, and specification of data management standards were all provided for in the MVP for the project. The project sponsor, EPM took it upon itself to establish a robust internal monitoring system based on regular assessment of, and reporting on, advances in both phases of the Institutional and Community Strengthening Plan (PFIC), which was essentially the Indigenous Peoples Plan for the project. This was furthermore complemented by the monitoring of visual impacts (which included culturally specific indicators), noise from the operation of the windmills, and water quality of the water coming out of the Kasiwolin desalination plant as part of Jepirachi's environmental monitoring protocol.

Designing Suitable Indicators. The generation of appropriate indicators to measure socioeconomic and, where appropriate, cultural, variables relating to a wind power project is typically the most challenging part of the design of any socioeconomic monitoring protocol. In this regard, a distinction should be made between indicators that are meant to measure the ultimate social development impacts of the project, and those meant to measure the outputs and outcomes of the implementation of the project's community engagement, compensation, and benefits-sharing activities, which is the basic goal of post-construction socioeconomic monitoring. Such impact and output indicators should be limited in number, be as easily verified as possible, and involve the collection of both qualitative and quantitative data. These indicators are effective when they are able to both: (i) gauge the quality of performance of the project sponsor or other entity charged with implementation of the community engagement and benefits-sharing strategy; and (ii) promote accountability for doing this within the relevant divisions of the project agency and among other relevant stakeholder groups, such as households that have entered into land leasing agreements.

Notes

1. When benefits-sharing arrangements allow local residents to have a direct economic stake in the development and operation of a wind project, the project's local image can seem even more enhanced. As a local Danish wind turbine owner succinctly put it: "A wind turbine from which I make a cent every time the rotor makes a revolution is so much prettier to look at, than one owned by someone else."

2. Kuran (2004) provides a succinct description of this approach to culture: "A community's culture consists of the beliefs, preferences, and behaviors of its members, along with the mechanisms that link those traits to one another. These traits give the community a unique identity that distinguishes it from other communities. This identity is subject to change, for a culture is a living organism. Through their interactions and their reactions to external influences, the members of a community transform their behaviors and also ultimately the underlying beliefs and preferences. By the definition adopted here, such changes amount to cultural change."

3. Lease payments may be referred to as "land rent," "profit sharing," "cooperative ownership," or "carried interest," depending on the particularities of the socio-cultural context and the nature of the benefits-sharing arrangements negotiated between the wind project developer and local residents.

4. Note that these figures refer to the total land area needed for the establishment of a wind farm, not to the more limited amount of land that needs to be acquired and cleared for civil works construction.

5. In fact, all the examples cited in this subsection are from projects that are currently receiving, or at one time received, support from the IFC. The details provided on the social management aspects of the projects are available on the IFC website (www.ifc.org/projects).

6. Due to historical and highly context-specific circumstances, there is no universal consensus on what makes somebody "indigenous." However, there is enough agreement on key characteristics of indigenous communities (such as their unusually strong attachment to the lands on which they live and the natural resources on which they depend) as to provide reliable guidelines for identifying them in most operational settings. Depending on the country, "indigenous peoples" can be referred to by such terms as "traditional peoples," "indigenous ethnic minorities," "minority nationalities," "tribal groups," "scheduled tribes," "hill tribes," or "aboriginals."

7. The UN Declaration was originally approved by 143 countries around the world, including all of those in Latin America and the Caribbean except for Colombia, which abstained, but later endorsed the document, and a handful of Caribbean nations (which were absent at the time of the vote).

8. For the Wayuu, the wind is sacred, and "Jepirachi" itself means "winds coming from the northeast" in Wayuunaiki, their native language.

9. The World Bank's safeguard policy on Involuntary Resettlement, OP 4.12, contains a technical definition of replacement cost (see the first footnote of Annex A). The main reason for the incorporation of this standard into the policy was so that landowners or occupiers who were poor or otherwise vulnerable would not suffer undue harm from the displacement that tends to result from expropriation.

10. Where indigenous peoples are among those staking a claim to the lands being acquired for a World Banksupported wind project, para. 17(b) of the Indigenous Peoples Safeguard Policy provides for the legal recognition of such a claim, based on evidence of the claimants' traditional occupation and usage of the lands in question.

CHAPTER 5

Final Considerations

Making a "Green" Technology even Greener. Wind power has emerged as a rapidly growing power generation technology with important climate benefits. Given the widely recognized need to find increased sources of low-carbon electricity, wind power is poised for continued rapid growth worldwide. Yet along with this growth comes the realization that wind power development poses its own particular set of environmental and social issues, as do all large-scale power generation technologies to varying degrees.[1] Consequently, as with any type of infrastructure or energy development project, it is important to acknowledge, assess, and mitigate any significant adverse environmental and social impacts that may arise during project planning, construction, and operation.

Wind-Specific and Broader Environmental and Social Impacts. This report has identified the diverse range of environmental and social impacts that tend to accompany land-based wind power development; the main impacts are detailed in Table 5.1. Some of these impacts are particular to wind turbines and include bird and bat collisions, visual impacts (including shadow flicker), and radar and telecommunications interference. Other impacts are associated with power transmission lines, which are needed to deliver wind-generated as well as other types of electricity to consumers; these impacts can also involve bird mortality and visual impacts, as well as the taking of land needed for transmission corridors. Still other impacts have to do with many kinds of large-scale civil works, including but not limited to wind farms. This last category of impacts can be negative or positive, and includes land clearing to install wind turbines and associated facilities, land acquisition leading to a loss of assets (even though preexisting land uses can often continue), increased local income and employment generation, cultural impacts on indigenous and other traditional rural populations, effects on physical cultural resources, disturbance of wildlife and sensitive ecosystems from the activities of construction workers and wind farm personnel, and the wide range of direct and induced impacts related to access road construction and improvement.

Feasible Mitigation and Enhancement Measures. For each of the environmental and social impacts identified, this report discusses feasible measures that can be used to mitigate the adverse effects—and enhance the positive ones—of wind power development. These suggested measures are based on existing technologies and good practices that are already in use to varying degrees. Among the key mitigation and enhancement measures, careful site selection stands out as particularly important. Furthermore, among the range of potential sites with highly favorable wind conditions and adequate proximity to transmission lines, choosing the lower-risk sites—in terms of biodiversity, local nuisances, and socioeconomic and cultural impacts—will reduce the need for potentially costly mitigation measures and improve environmental and social outcomes in general. Another major consideration is effective stakeholder engagement, which is important for addressing a broad spectrum of environmental and social concerns—from visual impacts to compensation and benefits-sharing. This report presents a variety of planning tools that can be used to improve site selection, engage effectively with stakeholders, optimize project design, and select mitigation and enhancement measures. Some mitigation and enhancement measures typically involve multiple objectives and thus are best selected with these tradeoffs in mind; examples include wind farm landscape management and decisions about public access. Other mitigation or enhancement measures are aimed at addressing specific issues, such as compensation for land taken, benefits-sharing with local communities, post-construction bird and bat monitoring, increased turbine cut-in speeds, short-term turbine shutdowns, environmental rules for contractors and construction workers, and conservation offsets. The extent to which many environmental and social mitigation or enhancement measures are implemented will often be influenced by cost-effectiveness or other financial or economic considerations.

Learning Curve. This report discusses feasible measures for managing environmental and social impacts that are based on existing knowledge and currently available technologies. For some types of impacts, such as bird and bat mortality at wind turbines, many gaps remain in scientific knowledge; as this knowledge improves over time, mitigation measures can be further optimized. There is also room for innovation and learning regarding how to optimize benefits-sharing arrangements and other measures for addressing socioeconomic and cultural issues in wind power development. Another important variable is technological change in turbine design and other advances in the wind power industry. As learning improves and technologies change, the preferred environmental and social mitigation and enhancement measures will also change over time. In this regard, this report describes environmental and social good practice measures that are available today and already in use. Appropriate use of these measures will help to ensure that wind power maintains its positive environmental image and fulfills its potential as an environmentally and socially sustainable energy source.

Table 5.1: Environmental and Social Impacts of Wind Power Projects and Corresponding Mitigation or Enhancement Options

Impacts	Project Mitigation/Enhancement Options		
	Planning	Construction	Operation and Maintenance
Biodiversity Impacts			
Bird Mortality: Birds collide with spinning wind turbines; also meteorological towers with guy wires and power transmission lines. Bird species groups of special concern include raptors, seabirds, migratory species, and birds with aerial flight displays.	Careful **site selection** of wind farms and associated transmission lines to favor lower-risk sites (such as many cultivated lands, non-native pastures, and deserts away from oases), while seeking to avoid higher-risk sites (such as many shorelines, wetlands, small islands, migration corridors, and designated Important Bird Areas). Within a planned wind farm, adjusting the location of turbine rows or individual turbines can further reduce adverse impacts. Consider choosing **wind power equipment** that is relatively more bird friendly (where consistent with other objectives), particularly in terms of larger turbine size, reduced or different night lighting, minimal number of guyed meteorological towers, power poles with bird-friendly configurations, and transmission lines with bird flight diverters in higher-risk areas.		Use **post-construction monitoring** to: (i) verify actual bird impacts; (ii) enable adaptive management; and (iii) assess potential impacts of scaling up wind development in the same general area. Consider **short-term shutdowns** (feathering) of turbines during peak bird migration events. Also perform diligent **equipment maintenance** to ensure that holes in turbine nacelles are capped to prevent bird entry.
Bat Mortality: This occurs when bats collide with spinning turbines or closely approach them, causing lung damage from decompression. Bat mortality typically exceeds bird mortality at wind turbines.	Careful **site selection** of wind farms to avoid higher-risk sites (such as wetlands, wooded areas, known migration corridors, and near caves).		Use **post-construction monitoring** to: (i) verify actual bat impacts, (ii) enable adaptive management, and (iii) predict potential impacts of scaling up wind development in the same general area. Consider operating wind turbines at an **increased cut-in speed** (such as 6 m/sec instead of 3-4), which can substantially reduce bat mortality with relatively minor losses in power generation.

Impacts	Project Mitigation/Enhancement Options		
	Planning	Construction	Operation and Maintenance
Wildlife Displacement: Some open-country birds such as prairie grouse are displaced from otherwise suitable habitat because they instinctively stay away from tall structures, including wind turbines and transmission towers. Large, shy wild mammals such as antelopes can also be displaced by the regular presence of wind farm employees.	Careful **site selection** of wind farms and associated transmission lines to avoid critical habitats and concentration areas for these sensitive species.		
Natural Habitat Loss and Degradation: Establishing rows of wind turbines with interconnecting roads can involve the clearing and fragmentation of natural habitats, sometimes affecting scarce ecosystems such as mountain ridge-top forests. Semi-arid ecosystems can also be degraded by careless off-road driving by wind farm personnel.	Careful **site selection** of wind farms and associated transmission lines to: (i) avoid existing and proposed protected areas and other sites of high conservation value and (ii) avoid or minimize the clearing and fragmentation of natural habitats in general, particularly native forests. Configuring wind turbines and access roads to minimize the clearing and fragmentation of natural habitats. With respect to **wind power equipment**, overall land clearing (on a per-MW basis) will be less with a smaller number of larger turbines.	Enforcing **environmental rules** for contractors that require, among others, minimum clearing of natural vegetation for all wind farm facilities; no washing of machinery or other pollution of waterways or wetlands; and no hunting, off-road driving, or other needless disturbance to natural habitats by construction workers.	Ensuring that wind farm personnel follow **environmental rules** of conduct (as during construction).
Overall Biodiversity Impacts: This includes bird and bat mortality, wildlife displacement, and/or natural habitat loss and fragmentation.	Use **planning tools** to optimize project site selection and design; these tools can include strategic environmental assessments (SEAs, including cumulative impact assessments); project-specific environmental impact assessments (EIAs), including environmental management plans (EMPs) and sometimes pre-construction biodiversity studies; overlay maps; and zoning maps. Support **conservation offsets** to protect and manage high-quality habitats, or for other enhanced management of the species of concern, away from the wind project area.		Use **landscape management** at wind farms to achieve desired objectives, which could include: (i) maintaining preexisting land uses; (ii) conserving and restoring natural habitats; (iii) managing land for species of conservation interest; (iv) deterring bird or bat use, as a means of reducing mortality; and (v) facilitating bird and bat monitoring. Manage **public access** as needed to protect vulnerable species and ecosystems, along with other objectives (where relevant) such as: (i)maintaining previous land uses; (ii) ensuring public safety; (iii) minimizing the risk of sabotage or theft of wind power equipment; and (iv) promoting local tourism and recreation.

Impacts	Project Mitigation/Enhancement Options		
	Planning	Construction	Operation and Maintenance
Local Nuisance Impacts			
Visual Impacts: Some people consider large wind turbines to be an eyesore. Local concerns about the visual impacts of wind turbines and/or their associated transmission lines can be a constraint to wind power development, particularly in developed countries.	Thorough **stakeholder engagement**, including prior consultation, participatory decision-making, and information disclosure and dissemination. Careful **site selection** of wind farms and associated transmission lines, with special attention in areas where local stakeholder sensitivities may be high, including sites importance for tourism and recreation. Within a planned wind farm, adjusting the location of turbine rows or individual turbines to reduce perceived visual impacts. Choose **wind power equipment** with aesthetics in mind (where consistent with other objectives), such as a smaller number of (larger) turbines and reduced or different night lighting. Use **planning tools** to: (i) assess specific visual impacts (viewshed mapping, photo composition, virtual simulations, and field inventories of views); and (ii) include visual impacts among other considerations within SEAs, EIAs, overlay maps, and zoning maps.		
Shadow Flicker: This is a specialized type of visual impact, in which spinning wind turbines create an annoying effect of rapidly blinking shadows when the sun is near the horizon.	Careful **site selection** to locate turbines where they would not produce shadow flicker around human dwellings. Use **planning tools**—standard industry software that predicts the location and timing of shadow flicker. **Stakeholder engagement** with potentially affected households and businesses.	Planting trees to provide a **visual screen.**	Consider **short-term shutdowns** of individual turbines during the brief periods when shadow flicker would affect dwellings.
Noise: Wind turbines produce both mechanical noise (turbine hum) and aerodynamic noise (rotor swish), which humans readily notice within 300 m or more.	Careful **site selection** to locate turbines an adequate distance from human dwellings. Use **planning tools**—standard industry software that predicts specific noise impacts on nearby buildings. **Stakeholder engagement** with potentially affected households and businesses.		

Impacts	Project Mitigation/Enhancement Options		
	Planning	Construction	Operation and Maintenance
Radar and Telecommunications Interference: Operating wind turbines can interfere with the signals received by radar and telecommunications systems, including aviation radar, radio, television, and microwave transmission. These impacts tend to be significant when wind turbines are within the line-of-sight of the radar or telecommunications facility.	Careful **site selection** to avoid installing turbines within the line-of-sight of radar or telecommunications facilities. Use **planning tools** to assess these impacts within SEAs, EIAs, overlay maps, and zoning maps. For turbines within the line-of-sight of aviation radar, consider **additional investments** to relocate affected radar, blank out the affected radar area, or use alternative radar systems to cover the affected area.		
Aircraft Safety: Wind turbines can pose a risk to aircraft if located too close to airport runways. In agricultural areas, the presence of turbines will preclude the aerial spraying of crops.	Careful **site selection** to maintain adequate distance between airport runways and wind power facilities. Use **planning tools** to assess these impacts within EIAs, overlay maps, and zoning maps. Where relevant, **stakeholder engagement** with agricultural interests.		In agricultural areas, use **alternative systems**, such as ground-level spraying or a shift to organic production.
Blade and Ice Throw: There is a very small risk of a loose rotor blade being thrown as a result of severe mechanical failure. In cold climates, there is also a small risk of rotor blades throwing off chunks of ice when they begin to rotate.	Careful **site selection** to locate turbines an adequate distance from human dwellings.		
Socioeconomic and Cultural Impacts			
Land Acquisition Involving Displacement: Although there are different means of acquiring land for a wind farm, formal expropriation involving negative impacts on land and land-based assets is sometimes necessary.	Careful **site selection** to locate turbines and associated infrastructure in such a way as to avoid or minimize the need for expropriation and related physical and/or economic displacement. Use of **participatory planning tools**, such as strategic environmental assessments that incorporate attention to social issues (SESAs), to optimize social values in project site selection and design. At the project level, these tools can include social assessments (SAs) and project-specific social impact mitigation plans.	Where expropriation and displacement are involved, specific measures are needed to ensure that affected people do not become worse off as a result of any displacement stemming from a project. Lost assets can be replaced through the **payment of compensation** and/or the provision of a variety of **local benefits.**	In cases of significant adverse impacts on assets, conduct an **ex-post evaluation** to confirm that those assets have been successfully replaced.

Impacts	Project Mitigation/Enhancement Options		
	Planning	Construction	Operation and Maintenance
Livelihoods and Income: Because wind power development has shown potential for generating a number of localized social benefits, the impacts on the people involved have tended to be positive.	Implementation of inclusive **stakeholder engagement** strategies. Use of the results of social analytic processes (such as SA) to develop **benefits-sharing arrangements** based on one or more of the following: (i) payment of rents or royalties, (ii) clarification of property rights, (iii) generation of employment, (iv) implementation of local benefit programs, and (v) promotion of community ownership of wind farms.	Use of **local labor** in wind farm construction.	Use **socioeconomic monitoring** to confirm that promised social benefits are delivered throughout the life of the project. Stakeholder consultation on, and coordination of, **landscape management** actions. Use of qualified local labor to assist with maintenance and security of wind turbine facilities.
Indigenous Peoples: While indigenous peoples stand to benefit from wind power projects in various ways, their vulnerability can make wind power development on their lands particularly challenging.	Implementation of **culturally differentiated engagement strategies**. Use of the results of social analytic processes (such as SA) to develop an **Indigenous Peoples Plan** (IPP) that provides for both the mitigation of any adverse impacts and the delivery of culturally compatible benefits to the communities involved.		Use **socioeconomic monitoring** to confirm that culturally compatible benefits are delivered to the communities involved throughout the life of the project.
Physical Cultural Resources: Proximity of wind farms to areas where physical cultural resources (such as archaeological or paleontological remains) are known or suspected to exist can result in negative impacts on them.	Use of **project-specific EIA** to assess the probability of impacts on physical cultural resources. In cases of confirmed presence of such resources, incorporation of specific **protection measures** into a project's EMP.	Where physical cultural resources are unearthed during project construction, use of **chance finds procedures** that are part of the EMP and the rules for contractors for a project.	Use **post-construction monitoring** to confirm that any continuing impacts on physical cultural resources are being adequately managed.

Notes

1. Table 2.1 provides an illustrative listing of some of the diverse environmental and social impacts often associated with different power generation technologies.

Appendixes

Appendix A. Case Study: Mexico La Venta II Wind Power Project

Project Description

The World Bank–assisted Mexico Wind Umbrella Carbon Finance Project supports the development of the La Venta II wind farm, located in the Ejido La Venta, in the municipality of Juchitán de Zaragoza, in the state of Oaxaca. The project consists of a wind farm of 98 turbines with a total nominal capacity of 85 MW, along with the associated interconnection system. The wind farm provides annually about 340 gigawatt-hours (GWh) on average, thereby reducing the emission of some 200,000 tons of CO_2 (tCO2e) from fossil fuel-based power generation. The plant is owned and operated by the Mexican Government's Federal Electricity Commission (Comisión Federal de Electricidad, or CFE) and is the first large-scale wind power plant to become operational in Mexico. The environmental and social findings reported here are based largely on the Study Team's October 2008 field visit to the project area.

Main Environmental and Social Issues

The main environmental issue associated with the La Venta II wind farm is the potential impact on birds and bats, particularly due to collisions with the wind turbine rotors. As was recognized during project preparation, the La Venta II wind farm site—along with the broader wind resource area in the southern part of Mexico's Isthmus of Tehuantepec—is a world-class bird migration corridor, through which millions of birds pass each year as they move between North America and Central or South America. Some of the key findings from recent bird and bat monitoring at La Venta II are:

- Over 90 percent of the entire world population of Swainson's Hawk *Buteo swainsonii* and Franklin's Gull *Larus pipixcan* passes seasonally through the south Tehuantepec wind resource area that includes La Venta II, as well as many other wind farms under construction or planned. No carcasses had yet been found of either of these species, although the short-term shutdown procedure described below remains an important safeguard for special circumstances, such as large flocks passing through during unusual weather.
- The most highly localized species occurring at La Venta II is the Cinnamon-tailed Sparrow *Aimophila sumichrastii*, which is found in the low-growing thorn forest of the south Tehuantepec wind resource area and nowhere else on the planet. Monitoring to date indicates that this species remains common within suitable habitat at La Venta II, even though a small fraction of its remaining habitat here was cleared for wind farm platforms and access roads, and there is some turbine-related mortality (1-3 carcasses were found in 2007).
- During most most of 2007 and early 2008, a total of 78 bird and 123 bat carcasses were found at La Venta II. However, the actual mortality of bats and small birds is undoubtedly many times higher (perhaps by a factor of 50 or so), because of: (i) rapid removal of small carcasses by scavenging animals, between the once-weekly searches; (ii) the inability to search much of the target area due to the type of vegetation and/or not enough searchers; and (iii) the tendency to overlook small, mostly camouflaged carcasses, even in open fields. For large

bird carcasses (raptors and large water birds), the correction factor between observed and actual mortality should be much smaller because the carcasses tend to be scavenged on-site so that the physical evidence lasts for a longer period, and are less likely to be overlooked during a standard search.

■ For local bat species, it is possible that La Venta II is functioning as a "population sink," a situation in which mortality exceeds reproduction and the local population is maintained through influxes from adjacent areas. However, none of the 19 species found dead during post-construction monitoring through 2008 to date are considered threatened, and none are highly localized in their distribution.

The one bird species which seems to be significantly affected around La Venta II is the White-tailed Hawk *Buteo albicaudatus*, a non-migratory species with foraging behavior—flying mostly at rotor-swept area (RSA) height—which puts it at high risk of colliding with wind turbines. Post-construction monitoring found 7 carcasses under turbines during the 2007-2008 monitoring period (INECOL 2009). When the cumulative impacts of multiple wind farms are considered, it appears that the south Tehuantepec Wind Resource Area (WRA) might become a local population sink for this species. This species is classified as "Subject to Special Protection" under Mexico's threatened species regulations (NOM-059). However, it remains reasonably common elsewhere in certain grassy lowlands of Mexico and Central and South America.

To date, other bird mortality at La Venta II does not appear significant enough to have measurably affected local populations. However, there is concern about the potential cumulative impacts of the many additional wind farms planned in the same general area, especially with respect to birds that migrate at night.

Regarding social issues, the land used by residents of the local *ejido* was adversely affected by the construction of the platforms for the installation of turbines and other related infrastructure. In accordance with the Indigenous Peoples Plan developed by CFE, and a Social Benefits Plan or *Acta Social* agreed with the *ejidatarios*, compensatory measures based both on household-specific and collective lease payments, as well as on the provision of community infrastructure, were undertaken. These are described in greater detail below. The Zapotec indigenous heritage of the residents of Ejido La Venta was documented in the project EA report, which incorporated demographic data and a complementary discussion of socio-cultural issues in the project area.

Mitigation or Enhancement Measures Taken or Proposed

INECOL's ongoing bird and bat monitoring activities include: (i) daytime observation of migrating bird flocks, their numbers (by species), and the routes they follow; (ii) use of a mobile ship radar (mounted on the back of a pickup truck) to detect migratory bird flocks by day and night; (iii) systematic observations of how birds behave when flying through the wind farm (including turbine avoidance reactions); (iv) monitoring of local bat and non-migratory bird populations, including their habitat use within the wind farm, bird nesting activity, and searches for caves and other bat roosting areas; and (v) periodic searching for and removal of bird and bat carcasses around the turbines.

CFE and INECOL have agreed upon an innovative and highly effective procedure for real-time, short-term shutdowns of wind turbines when large flocks of birds are detected, by radar and/or visual observation, flying towards the wind farm at or near RSA height. This procedure involves INECOL's bird spotters notifying CFE technicians in the wind farm's control room via cell phone when a large flock is approaching around RSA height, so that the CFE technicians can immediately cause the rotor blades of all the turbines—or those in a particular at-risk portion—of the wind farm to be feathered (so the rotors stop spinning), until the bird flock has passed by the wind farm. Since they began work in the autumn of 2007 until spring 2009, INECOL monitors have requested only three such brief shutdowns (not counting a few rehearsals) because most large bird flocks pass over the wind farm well above the maximum RSA height of 72 m. The three requested shutdowns were for brief periods (well under one hour each) and implied negligible reductions in La Venta II's overall electricity generation. However, under unusual weather conditions, these large flocks do tend to fly through La Venta II at a lower altitude that places them at significant collision risk. (This phenomenon was observed during the EA process in 2005, before La Venta II became operational.) Thus, the procedure for short-term turbine shutdowns at La Venta II remains available as an infrequently used, though nonetheless important, contingency measure.

One specific problem—with a simple solution—noted during the World Bank mission's visit was that some wind turbine nacelles (gondolas) had round holes because their covers were missing, and that American Kestrels *Falco sparverius* were using these holes for shelter or attempted nesting, but with fatal consequences because of the close proximity of the rotors. Five kestrel carcasses had been found beneath four such turbines. CFE informed the World Bank that their personnel would soon enter the nacelles to cover up all such holes.

With respect to indigenous peoples, *ejidatarios* whose lands were affected by the construction of platforms for the installation of turbines and related infrastructure have been receiving fixed annual payments (using the *"cuota fija"* modality). Levels of reported satisfaction with the amounts involved have been generally high. In fact, the social benefits provided by the project have been so apparently attractive, they have spurred a group of *ejidatarios* who have land holdings within La Venta II's area of influence but who (for various reasons) had opted not to participate in the project when it was being developed, to submit a formal petition to CFE for inclusion. After evaluating the wind farm expansion needs implied by the petition, the Commission decided that it did not have sufficient financial resources to accept it. CFE agreed that it would explore appropriate ways to communicate this decision to the small number of *ejidatarios* involved, in an attempt to maintain the generally good community relations that have characterized this project. Although this sub-group of landholders has not been receiving annual fixed payments, they benefit from communal land use payments from CFE. In addition, the Oaxaca state government is considering developing productive projects for the affected landholders.

Some Lessons

The World Bank observed that not much has been done to reflect on experiences and disseminate lessons in the management of social issues in La Venta II, even though examples abound. Many of these are apparently being applied in the preparation by CFE of La Venta III, to be sited in Ejido Santo Domingo. CFE and the World Bank agreed on the need for such an effort.

The La Venta II Project demonstrates that a specific procedure for short-term turbine shutdowns (feathering of rotor blades) in real time during peak bird migration is technically feasible and can be highly cost-effective.

The low-growing native thorn forest of the southern Isthmus of Tehuantepec WRA is the world's only habitat for the Cinnamon-tailed Sparrow; it also harbors other animal and plant species of conservation interest. None of this habitat presently is within any type of protected area. Wind power projects located within this WRA could make a very positive contribution to biodiversity conservation by working out binding agreements with landowners to maintain intact this native vegetation, and adjusting accordingly the amounts paid to landowners for hosting the wind turbines.

Appendix B. Case Study: Colombia Jepirachi Wind Power Project

Project Description

Jepirachi is a wind power project of 19.5 MW produced by 15 turbines of approximately 60 m hub height. The project, which in 2005 produced 49,358 megawatt-hours (MWh) of power and generated net revenue of Col\$3.741 million (US\$1.5 million), is the first grid-connected wind power project in Colombia. It is located on the Guajira Peninsula, in the northernmost part of Colombia very close to the Venezuelan border. The area is inhabited by the Wayuu, one of the more numerous of Colombia's 80 indigenous groups. The Wayuu economy is based mostly on fishing, goat herding and, to a lesser extent, horticulture.

The Jepirachi project is expected to displace an estimated 430,000 metric tons of carbon dioxide (CO_2) until 2019. Moreover, through the supply of Emissions Reductions (ERs) as developed under the Clean Development Mechanism (CDM), the project has been facilitating the development of a domestic carbon market in Colombia.

The hot, arid region where Jepirachi is located has considerable wind energy potential. A 200 MW wind farm has been under study for another part of the Wayuu homeland to the west. Jepirachi therefore has pilot project character for potential larger-scale wind power development in the Guajira Peninsula.

The Jepirachi project was built by Empresas Públicas de Medellín (EPM), a large public corporation with a varied portfolio in infrastructure. Concerning power generation, EPM has in recent years focused on building large hydropower plants in various parts of the country as well as in neighboring countries (such as the Bonyic project in Panama). Jepirachi is the first wind power project built by EPM, and as such represents an attempt by the company to diversify its generation options by learning about new generation technologies.

Obtaining external financing (in this case, through the Bank's Carbon Finance Unit) for the avoided emissions that result from relying on low-carbon technologies like wind power means that EPM is learning to use emerging international carbon markets to finance eligible power projects. Emissions monitoring is provided by an independent verifier, DNV (Det Norske Veritas).

Main Environmental & Social Issues and Mitigation or Enhancement Measures Taken or Proposed

A sound Environmental Management Plan (EMP) has been prepared and is being followed. This was seen, for example, in the protection of vegetation in the vicinity of the wind turbines, in that care was taken both during construction and subsequently through replanting of areas that had to be cleared for the construction. EPM monitoring reports are critical of current practice where necessary, recommending needed improvements in response to the issue at hand (such as water quality from the desalination plant).

Good environmental practice was followed during construction. In particular, the archaeological recovery efforts employed went above and beyond the requirements of the cultural property safeguard policy (see the discussion on OPN 11.03 under "Social Issues" below). The World Bank mission also made favorable note of the handling of waste resulting from the project in a sanitary landfill in Uribia.[1] Hazardous wastes (for example, lubricants) are even transported back to Medellín for proper disposal as EPM staff believed that this could not be adequately done locally.

Wind power is a very promising and rapidly growing renewable and carbon-neutral technology for generating electricity. However, there is growing evidence worldwide that some wind farms pose a significant problem for birds and/or bats, either directly through collisions with the rotor blades, or indirectly through displacement from otherwise suitable habitat. During project planning, Jepirachi was considered to be a site of relatively low risk for bird collisions, as indicated in the Environmental Assessment (EA) report. The site appears to be of relatively low risk for bird mortality, due in part to its being intentionally set back from the coast. Also, because of the sparse, semi-arid vegetation and general lack of fresh water, the Jepirachi area has a relatively low density of resident, land-based birds that routinely fly high enough to be in range of the rotors. A World Bank safeguards supervision mission in August 2007 reviewed the issue of bird collisions and associated mitigation and monitoring efforts at Jepirachi, along with other environmental and social concerns.

The World Bank mission found that the bird mortality monitoring carried out at Jepirachi involved a rather low level of effort in terms of field-days of monitoring and had not produced reliable written data. For these reasons, bird mortality from turbine collisions at Jepirachi could not be quantified, in terms of estimated collisions/turbine/year, or by species. Nonetheless, bird mortality at the wind farm does not appear to be unduly high overall, nor concentrated among a particular species. In searching around the base of most of the turbines, a brief World Bank visit in 2007 found no carcasses of dead birds (which, over time, tend to be removed by scavenging animals.) However, Jepirachi staff as well as local residents reported having previously found the mutilated carcasses of various larger birds around the turbines, a clear indication that they had collided with the rotors. With the exception of the distinctive Brown Pelican *Pelecanus occidentalis,* the bird descriptions provided by these witnesses were too imprecise for positive identification by species. During its two days at Jepirachi, the World Bank observed several Turkey Vultures *Cathartes aura* and a White-tailed Hawk *Buteo albicaudatus* engaging in "risky behavior," that is, flying 30-90 m high near a turbine.

The World Bank visit found no evidence of any bats around the turbines (or elsewhere in the Jepirachi area) at dusk and after nightfall, even with the use of a powerful flashlight. Nor did anyone interviewed by the World Bank mission report the presence of any bats, alive or dead. The apparent scarcity of bats around Jepirachi seems to be related to the lack of fresh water and the naturally sparse, low-growing vegetation.

With respect to nonlethal impacts, the World Bank mission observed that many birds (including localized arid zone endemics such as the Vermilion Cardinal *Cardinalis phoeniceus* seemed to be displaced from otherwise suitable habitat due to turbine noise for a radius of about 50 m from the base of each windmill. Such an impact would not be significant, especially since the semi-arid habitat found at Jepirachi still occurs extensively on the Guajira Peninsula.

The World Bank mission recommended that EPM improve its bird mortality monitoring and data management. Even though the World Bank found no evidence that bird mortality is at problematic levels, systematic monitoring is important because: (i) with the rapid worldwide expansion in wind power capacity, there is an urgent need to improve scientific knowledge in general, regarding which types of wind turbine placement, design, and operation tend to be the safest for birds and bats, and why, and (ii) plans exist for scaling up to a much larger wind farm (around 200 MW) on the Guajira Peninsula, and reliable data from Jepirachi would improve the environmental assessment, planning, and operation of subsequent facilities. Conducting scientifically credible bird monitoring—and transparently disclosing the data—would also be consistent with EPM's often-stated desire to be an environmental leader in the field of electric power generation.

The World Bank mission recommended that EPM adopt the following, relatively low-cost measures at Jepirachi to restore and maintain its bird monitoring program at a minimally adequate level:

- **Systematic Monitoring** A well-qualified, independent specialist (from a university, NGO, or consulting firm) with several field assistants should visit the wind farm periodically, to search systematically for bird or bat carcasses around each turbine and compile and organize the data, including any from chance finds by wind farm personnel. Each visit should involve several days of field work, to ensure that a sufficient radius (perhaps 100 m) around each wind turbine is adequately searched. A sample of turbines should be searched on consecutive days to help calibrate any loss of data due to carcass removal by scavenging animals. There should be a minimum of at least two such visits per year, preferably during the southbound (peaking in September) and northbound (peaking in April) bird migrations; ideally, there would be six (or more) visits annually, one each during the migration months of March, April, May, August, September, and October.

- **Chance Finds Procedures.** Security guards and all other field personnel at Jepirachi should be instructed to follow a simple, but consistent protocol upon finding any dead birds or bats in the vicinity of the wind turbines. This protocol should include taking at least two photos of the carcass at the site it was found with a portable digital camera. The carcass should be turned over so that both sides are photographed. Each photo should have a ruler (to indicate size) next to the carcass, along with some means of showing the correct date in the photo. These photos are needed for subsequent identification of the birds (or bats) by species. A standard "wildlife incident report" form should also be filled out to provide basic data such as the exact location of the find, in relation to a particular turbine. Ideally, the carcass would be placed in a plastic bag and frozen until the next visit from independent scientific monitors. Since a well-functioning freezer might not be available at the Jepirachi control room, the bird carcasses should be deposited at a designated site away from the wind turbine area in order to prevent additional mortality from scavenging birds flying towards the carcass through the turbines and mistaken double-counting of the same carcass by different Jepirachi personnel.

- **Data Management and Sharing.** At least once per year, the independent monitoring team should summarize the data obtained in a concise report. This report should be shared with the World Bank's project task team and also disclosed publicly by EPM, such as on their Web page.

As confirmed in interviews conducted with Wayuu community members, visual impacts and noise generated by the wind farm did not appear to be posing a problem to the local communities. For both, EPM has carried out proper monitoring. The noise impact of the project is well within the legal limits for Colombia. The field visit confirmed that noise from the turbines is only evident within 100 m and the nearest human dwelling is much farther away than that. In the discussion with residents, they reported noticing noise only when the turbines started after having been previously shut down, typically for maintenance.

In the August 2007 supervision mission, World Bank representatives were impressed overall with the strides the project sponsor, with Bank support, had made with respect to promoting culturally appropriate development for Wayuu community members living in the indigenous *resguardo* (reserve) where the wind farm is located. The mission found that EPM's performance not only met the requirements of the Bank's former safeguard policy on indigenous peoples, OD 4.20 (since replaced by OP 4.10), but also surpassed them in many key areas related to project preparation and implementation.

Owing in part to thorough community-level consultations during the design phase of the project, EPM managed to avoid lopsided impacts and jealous reactions among rival clans in the project influence area by placing the two rows of wind turbines along either side of the border between the two main *rancherías* in the area, Kasiwolin and Arutkajui. The placement of turbines along each row varied slightly according to a few other criteria:

- Proximity to cemeteries, considered important stakes in territorial claims by different Wayuu clans;
- Proximity to areas of cultivation and the possibility that the project might obstruct wind getting to the areas of cultivation (given local beliefs that the wind plays a role in fertilization of the land); and
- Proximity to key archaeological sites (notably traditional Wayuu fireplaces present in the area).

EPM's commitment to the continued viability of the Jepirachi social action program (PSA) is shown by the incremental budget allocations it has made: these have totaled a multiple of the US$0.50 per ton of CO_2 reduced premium that the company is receiving for its implementation of the program. EPM has engaged in regular monitoring and evaluation of advances in the PSA.

EPM's main institutional counterpart on the Wayuu side, the Anna Watta Kai ("Well-Being for the Future") Foundation, has grown in size and influence in the sector of the *resguardo* where the project is located. With increased prominence has come increasing self-confidence and ambition, which is positive as long as there are mechanisms in place to keep the Foundation's leadership accountable to other community members. Interviews in the field suggested this may be an area of emerging problems that needs to be monitored.

The handling and preservation of physical resources having significant cultural and/or religious value for the Wayuu have been carried out in a manner that is in full compliance with the World Bank's Safeguard Policy on Physical Cultural Resources (OPN 11.03, now OP 4.11). Siting of all project components was chosen such that no locations of significance to the Wayuu were affected. The social benefits plan gives a detailed account of areas where there was a potential impact on local cultural patrimony and how this was handled. Over a period of six months during the project preparation phase, a good amount of relevant search, documentation, and preservation work was carried out.

Concerning the implementation stage, EPM has gone beyond supporting active archaeological recovery work by turning the findings into an educational tool whenever possible. For instance, books and even a game based on archaeological finds in the area were developed and handed out to teachers and children at the Kamusuchiwo'u school in Media Luna.

Some Lessons

In terms of the experience so far with the provision of culturally compatible benefits, the World Bank offered a number of observations and recommendations:

- **Tour Guides.** Several Wayuu were trained to be *etnoinformadores* (tour guides) who could accompany visitors to the area. Even though at the time of the field visit it appeared that many of those who went through the training were no longer carrying out this activity, various other examples of employment of mainly young Wayuu in connection with the wind farm (security guards, drivers) were observed.
- **Kamusuchiwo'n School in Media Luna.** The school has an enrollment of 700 students in a bilingual (Spanish-*Wayunaiki*) program. A number of improvements were made to the physical plant, as well as to the curriculum. The improvements appeared sound and appropriate.
- **Health Post in Media Luna.** Attempts to provide electrification to a local health post turned out to be unsustainable. During project preparation, the Bank's position was that the health post should be powered by solar energy rather than by diesel (which had been recommended by EPM, due to the risk of a more materially valuable solar generator getting stolen). Indeed, the solar panel eventually installed was stolen after only a year of operation, leaving the health post without electricity. In hindsight, the Bank could have paid more attention to the argument for diesel by EPM. Moreover, once the decision to use solar energy had been taken, the Bank could have been more actively involved in targeted communications, community information meetings, and similar actions aimed at conveying to the Wayuu the importance of protecting the solar panel.
- **Desalinization plant in Kasiwolin.** One of the more substantial components of the initial social benefits plan involved the construction of a desalinization plant for the Wayuu. While the plant has successfully provided water to the community (five days per week to 230 people in private households plus to the above-mentioned school with 700 pupils), and appropriate monitoring measures were implemented to monitor water quality, the system has also experienced significant maintenance challenges. For example, at the time of the World Bank visit the desalinization plant had been out of order for about two months, with no remedy in sight. On the positive side, the Municipality of Uribia filled the breach by providing water to local residents—something which had not happened before, indicating that the project prompted the local authorities to begin taking their service delivery obligations more seriously.

EPM opted to carry out a strategy of long-term strategic engagement with the Wayuu, which was highly sensible considering that: (i) EPM did not have experience working with indigenous communities prior to Jepirachi, and (ii) engaging with indigenous peoples ought to involve longer time horizons. This is often due to the difficulties inherent in cross-cultural communications, although weak capacity within indigenous organizations can also play a role.

The World Bank learned of several situations where poor communications across cultural boundaries caused friction in the otherwise smooth relationship between EPM and the participating Wayuu. At least one time, confusion over the exact benefits to be delivered under the PSA, possibly exacerbated by incorrect translation, made it necessary for EPM officials to reconvene community members in the *resguardo* to make their intentions clear, with the help of a new translator.

In another example involving the World Bank, the Bank's project team had indicated publicly that it would be able to secure grant funding for an on-site sustainable electrification project. The announcement raised community members' expectations at a very early point in the process, and these residents eventually came to believe that what mattered above all was to obtain access to electricity with the help of the Bank. Local frustrations grew when the grant application process took much longer than expected. Even though the initial consultations had not revealed electrification to be a priority for the Wayuu, their newfound desires and heightened expectations led to a distinct loss of momentum in the implementation of other aspects of the social action program, as they halted a number of efforts (including the opening of a restaurant for tourists) while waiting for a decision on the Bank grant proposal. A key lesson would be to guard against raising expectations in similar situations, given that the spoken word carries great weight in many indigenous cultures, including the Wayuu. Another lesson points to the need for effective coordination among the Bank (as a source of both financial and technical assistance), the project sponsor, and the local government in the implementation of community-based development activities, after the most feasible options have been identified and confirmed with community inputs.

From a bird and bat conservation standpoint, the Jepirachi Wind Power project appeared to have been following good practices with respect to: (i) turbine placement—set back from the shoreline by a minimum distance of 200 m, which ideally would have been greater; (ii) turbine design—almost no structures on which birds could perch; and (iii) turbine operation—minimal night lighting. At the time of the Bank's visit, bird monitoring during wind farm operation was the main area needing improvement. An adequate program of bird monitoring and data sharing (along the lines mentioned above) could have been more clearly described in the project's technical and legal documents.

Notes

1. Use of a sanitary landfill is generally regarded as more environmentally benign than use of an open dump, since in the case of the former wastes are isolated from the surrounding soil through lining materials.

Appendix C. Case Study: Uruguay Wind Farm Project

Photo: UTE

Wind turbines (2 MW each), Sierra de Caracoles, Uruguay

Project Description

The Uruguay Wind Farm Project is a 10 MW wind power plant located in Sierra de los Caracoles. This project is the first of its kind, designed as a small-scale pilot project, aimed at being easily replicated if successful. With a total plant factor of about 38 percent, the wind farm produces about 33 GWh per year. The wind farm is comprised of five turbines, 2 MW each, located in a single line atop a hilly ridge within the Sierra de los Caracoles, about 20 km from the town of San Carlos in the Department of Maldonado.

The turbine generators are from Vestas, model V80. Each wind turbine generates electricity at 690 volts (V). The voltage is raised to 31.5 kilovolts (kV) using dry type transformers located in each wind turbine tower. In addition, for each generator, there is an electronic system to correct power factor, a dry type transformer 690/230 V for the auxiliary services and an interconnection pane with circuit breakers. The technology for connection to the grid was provided by UTE, Uruguay's national electricity company. The interconnection to the network has been realized through a 60 kV line (operating at 31.5 kV) of about 16 km in length and a boosting transformer.

The wind power project seeks to contribute to Uruguay's sustainable development, most notably by increasing power supply and reducing the imports of crude oil and its derivatives. The project also reduces greenhouse gas (GHG) emissions through grid-connected electricity generation from wind power. These emissions reductions arise from the displacement of fossil fuel-based electricity generation on the national grid. The project is expected to generate Certified Emissions Reductions (CER) and revenues through the selling of CER under the Clean Development Mechanism. The revenues obtained will contribute to eliminate the barriers that prevent the implementation of this project, among which the high cost of electricity generation from wind power due to supply shortage in wind turbines manufacturing and irregularity of wind patterns. The emission reduction credits generated through the project activity will help Annex I countries[1] meet their emission reduction obligations as agreed under the Kyoto Protocol.

Main Environmental and Social Issues

In terms of potential collisions with the wind turbines, the site is considered to be of relatively low risk for birds and of uncertain risk for bats (Rodriguez et al. 2009). The site is not a bird migration bottleneck because of its distance from the coast and lack of other topography that would concentrate migrating birds into a small area. The migratory Swainson's Hawk *Buteo swainsonii* passes through the area, but in a highly dispersed (not spatially concentrated) manner. There are no wetlands or significant bird concentrations in close proximity to the wind farm. Although the Sierra de Caracoles harbors some bird species of conservation concern (endemic or threatened), these are all either: (i) species of wetlands or tall grasslands that do not occur close to the wind farm (Straight-billed Reedhaunter *Limnoctites rectirostris*, Black-and-White Monjita *Xolmis dominicana*, Chestnut Seedeater *Sporophila cinnamomea*, Entre Rios Seedeater *S. zelichi*, and Saffron-cowled Blackbird *Xanthopsar flavus*), or (ii) species of the *monte* forest that stay fairly close to the ground and are not likely to collide with wind turbines (Mottled Piculet *Picumnus nebulosus* and Yellow Cardinal *Gubernatrix cristata*). These species are thus at very low to zero risk of being adversely affected by the project.

In the case of bats, the level of risk from the wind farm is not yet clear, but will be determined through systematic monitoring. In general, bats are considered more vulnerable than birds to collisions with wind turbines because, for still unknown reasons, they appear to be attracted to the spinning rotors. Of the 21 bat species known from Uruguay, four are definitely known to occur in the project area (White-bellied Myotis *Myotis albescens*, Blackish Myotis *M. riparius*, Vampire Bat *Desmodus robustus*, and Brazilian Free-tailed Bat *Tadarida brasiliensis*), but other species are also likely. During preliminary post-construction monitoring during 30 days over 9 visits, two dead Brazilian Free-tailed Bats, a widespread and relatively abundant species, were found on April 19, 2009. No other bats and no birds were found dead during that initial monitoring period (Rodriguez et al. 2009).

The project involved the improvement of 7 km of an existing dirt road leading to the top of the ridge where the wind turbines are located, along with 4 km of new road. This has helped those landowners living near the road—and who are not directly affected by the project—to have easier, quicker access to route 39, which runs between San Carlos and Aiguá. In an information dissemination workshop held in February 2008, local landowners asked for UTE's assistance in rehabilitating another short section of road that linked up with the project's main access road, and that would give some landowners (otherwise unaffected by the wind project) easier access to Minas and other cities on the western side of the wind farm. This road was improved to the standard required for transporting equipment during wind farm construction, but UTE considered road improvements to a higher standard to be beyond the scope of their operations.

Local employment as a benefit of the Caracoles project has been limited, something not uncommon in wind farms beyond Uruguay. Figures provided by UTE indicate that about 39 percent of the total labor needed for the project was from the immediate area, but this was limited to the construction phase. At the time of the Bank's last visit to the project, in March 2009, only one person from a neighboring town was working on site, as a security guard. Operation and maintenance services are currently being provided by Vestas, the wind turbine manufacturer for the project.

A key difference between the Sierra de los Caracoles wind farm and similar projects visited by the Bank project team in Latin America is that the Caracoles case involved outright expropriation of the land used for the installation of the windmills and the construction of the control center. Sections of three contiguous plots (Nos. 20224, 785 and 783, according to Department of Maldonado's 2nd Cadastral Section) totaling nearly 27 ha were expropriated atop the ridge. This expropriation was carried out via Executive Decree in August 2006. From the start, one of the two landowners involved demonstrated satisfaction with the entitlements offered to him for his losses. In addition, beyond compensation for lost sections of two plots, the agreement included payment of an easement for use of part of an access road to the wind farm. The total compensation amount has already been paid to this owner. The second owner (No. 783) was initially not satisfied with the proposed compensation amount, and the matter was referred to the courts for resolution following procedures specified in Uruguayan law. An agreement has reportedly been reached, with a new payment due to this owner.

A total of 56 plots have required easements for access to the posts for the transmission line linking the wind farm with the San Carlos transformer station. In these cases, even as the easement is imposed, compensation is not provided automatically; it has to be requested by the affected landowner via an administrative procedure based on a claim of damages or harm (*daños y prejuicios*) connected to the running of the line through the property. According to UTE's easements policy, once they are notified of the need for an easement, landowners have 15 days to object to UTE regarding the alignment for the transmission line or related infrastructure; they can also request the Uruguayan Court of Claims to intervene. Landowners can also request formal expropriation if more than 75 percent of their land is needed for the easement, or if the affected land or other assets can no longer feasibly maintain their previous economic use.

Compared to other renewable energy projects supported by Bank carbon funds in various parts of Latin America, the Sierra de los Caracoles wind farm has generated relatively few local benefits beyond the payment of compensation for direct impacts. In this regard, while municipal officials in San Carlos reportedly expressed some interest in developing the wind farm site as a local tourist attraction, the idea was shelved after opposition was expressed by some local landowners, who were concerned that such a move would disturb the bucolic serenity of the Sierra de los Caracoles area.

Mitigation or Enhancement Measures Taken or Proposed

The monitoring of bird and bat collisions with the wind turbine rotors is considered to be a key environmental mitigation measure for this project. Even though the wind farm site is presumed not to be of high risk (at least for birds), such monitoring is nonetheless needed to: (i) verify whether or not a significant problem exists, particularly in the case of bats; (ii) enable the potential adaptive management of wind farm operation to minimize bat or bird mortality; (iii) predict the likely impacts of scaling up wind power development, particularly the proposed future expansion at the Sierra de Caracoles of another 10 MW, but also in other areas of Uruguay with similar physical and vegetation characteristics; and (iv) advance scientific knowledge worldwide, in a field that presently faces a steep learning curve and would surely benefit from the Uruguayan data. UTE prepared and agreed with the World Bank on a bird and bat monitoring program, which collected initial data during 2009. The program provides for at least three years of post-construction monitoring of bird and bat mortality (searching for carcasses) and specifies details such as: (i) the appropriate surface area to search around each turbine; (ii) appropriate correction factors to estimate the difference between observed and actual bird and bat mortality, including area not covered, scavenger removal, and observer efficiency; (iii) procedures for removing and documenting any carcasses discovered; (iv) maintaining contacts with international experts; and (v) management and dissemination of monitoring data. Most notably, the program provides for adaptive management of wind farm operation through (i) operating the turbines at the standard cut-in speed of 4.0 meters per second (m/sec) during Year 1; (ii) experimenting with 6.0 m/sec (1/2 hour before sunset until sunrise) during Year 2, if the Year 1 monitoring finds (with correction factors) more than five dead bats/MW/year; and (iii) if bat fatalities drop significantly during Year 2, then continuing with 6.0 m/sec during Year 3 (Rodriguez et al. 2009).

In order to ensure good environmental practices had generally been followed during the project construction, the World Bank conducted a field visit during the construction phase and also received evidence and explanations from the UTE staff (with photos and videos). Construction took place between August 2008 and December 2008. Land clearing (affecting less than 27 ha of mostly grassland), as well as the widening of short access roads, did not seem to exceed the minimum area required for the transportation and installation of the large wind turbines (hub height of 67 m). In particular, Bank staff noted the efforts that had been made during construction to minimize the clearing of any *monte* forest, as well as the removal of relatively short sections of an old stone fence of some historical interest. This fence is believed to have been built during the nineteenth century, although it does not qualify as an official historical landmark because many similar such fences are found in Uruguay. The fence is the only physical cultural property that was encountered during project planning and construction.

At the time of the World Bank's visit to Caracoles, compensation for transmission line easements had been requested by 6 owners in relation to 7 of the 56 plots. Most of these plots were closer to the San Carlos station, where the transmission lines from different projects converge, than to the mountains. Such was the case of a 71-year-old woman of modest means who was interviewed by the World Bank. Her 11 ha plot, which is not her residence but which she has kept in her family for recreational purposes, abuts the transformer station and is crossed by three different lines. At the time of the interview, she had just initiated the compensation request for the easement for the Caracoles line, for which she reported receiving considerable support from UTE field staff. She also said she had a very favorable impression of the wind farm, and that she had gone up the mountain to see the site while the project was still under construction. Another interviewee, a well-off 48-year-old man with a larger holding dedicated to cattle ranching, was further along in the compensation request process. Unlike the first interviewee, he had hired a lawyer to help him prepare his petition. He was not happy with the results of the initial cadastral valuation of damages from the running of the line through his lands, and had requested a re-valuation, which UTE appraisers were preparing to carry out. At the same time, he did not report anything unusual in his dealings with the company, in terms of its notification of the easement and explanation of the procedures involved for requesting compensation.

For UTE to have access to the ridge-top where the wind farm was constructed, it had to improve a number of sections of existing dirt access roads. This introduced temporary impacts associated with the rehabilitation works, all of which had to be compensated. Any fences or other such assets that sustained damage (*daños supervinientes*), either as a result of roads rehabilitation or the installation of posts for the transmission line, were also duly compensated.

Prior to the construction of the main project works, an information dissemination workshop was organized specifically for the landowners who were identified by UTE as being directly affected. Following circulation of an invitation to the meeting by UTE's head of press relations, the meeting itself took place on February 20, 2008 at the Society for Rural and Industrial Development of Maldonado. Three Society officers and 18 concerned local residents were in attendance. This meeting does not appear to have offered the most complete treatment of compensation payable to people directly affected by expropriations, easements, or damages as a result of the project. Nor was the discussion of other project benefits particularly complete. Nevertheless, the slack on local engagement was generally taken up by UTE staff with the strongest field presence, namely appraisers and legal advisers who were charged with contacting all affected landowners personally, in order to notify them of impending impacts and provide them with a sense of the entitlements available. UTE officials planned to hold at least two additional consultation events focusing on the environmental and social aspects of the project, one in the vicinity of Sierra de los Caracoles and one in Montevideo.

One proposed mitigation measure is to ensure the nacelle holes are not missing caps. During its March 23, 2009 field visit, the World Bank team found that the nacelle (gondola) for Turbine No. 5 had two round holes where the caps were missing (the other four turbines had all their caps). These holes can attract birds (such as the locally common American Kestrel *Falco sparverius*) to roost or attempt nesting within the nacelles, which is hazardous to the birds because of the close proximity to the rotors. (The same problem, with the same bird species, was found in at the Mexico La Venta II wind farm.) A Vestas representative present at the wind farm indicated that these two caps would be replaced very shortly.

Another proposed mitigation measure is to re-vegetate the cleared land. Neither UTE nor the construction contractor had made any efforts to re-vegetate (or facilitate natural re-vegetation of) the land around each turbine where the original grassland vegetation and topsoil had been cleared away during turbine installation. As a result, most of the land around each turbine has remained almost completely bare, with minimal grass re-growth to date. The rather extensive bare earth around the turbines poses low or moderate erosion risks; it also implies a larger-than-needed ecological footprint for the project. On the other hand, the relative lack of vegetative cover will facilitate finding any bat or bird carcasses around the turbines. UTE staff also expressed some reluctance to invest in land restoration around the turbines, arguing that natural regeneration will be adequate once the rains return. As a compromise, UTE agreed that it will demarcate with posts the area of parking lots around each turbine, as well as the interconnecting road, to prevent driving on additional land in the area. The land that is thus protected from being driven upon will have an improved opportunity for natural grassland regeneration.

The project involved two solid waste issues. First, the construction contractor had left behind various pieces of metal and other solid waste, both within the wind farm itself and just outside its boundaries near the main entrance. UTE agreed to require the contractor to remove all the remaining waste that it had left behind, which was done shortly thereafter. Second, the roadside area just outside the wind farm entrance has a small informal dump of regular garbage, which, according to UTE staff, has been left behind by local people (not UTE staff or contractors). UTE agreed that, even though not of their making, this dump can leave a bad impression on wind farm visitors and thus negatively affect UTE's image. Accordingly, UTE agreed to remove all the garbage that has accumulated, but then to inform local people that keeping the site free of garbage is their responsibility (not UTE's).

Some Lessons

The Caracoles project serves as an example of where formal expropriations have been used successfully, if in a limited way, to acquire the land needed for the main part of wind turbine facilities.

Social benefits could have been better defined and elaborated, and a more proactive stance could have been taken toward consultations with affected people.

UTE's agreement to experiment with a higher turbine cut-in speed if bat mortality is found to be significant represents the first known such case in a developing country.

Notes

1. Annex I Parties to the United Nations Framework Convention on Climate Change include the industrialized countries that were members of the Organization for Economic Cooperation and Development (OECD) in 1992, plus economies in transition including the Russian Federation, the Baltic States, and several Central and Eastern European countries.

Appendix D. Correction Factors for Real versus Observed Bird and Bat Mortality

Photo: Lygeum

It is much easier to estimate the mortality of large birds, such as this Eurasian Griffon Vulture Gyps fulvus *at a wind farm in Spain, than for small birds or bats that are more difficult to find and are frequently removed by scavenging animals.*

It is now widely recognized that the actual number of birds or bats killed at wind farms is normally larger—sometimes much larger—than the number of carcasses found in the vicinity of the turbines (Gauthreaux 1995, Orloff and Flannery 1992, Anderson *et al.* 1999, Morrison *et al.* 2001, Erickson *et al.* 2002, Smallwood and Thelander 2004, Smallwood 2007, CEC 2008). These discrepancies between observed and actual mortality occur because of real-world influences such as: (i) impenetrable vegetation or otherwise inaccessible terrain within the search area around a turbine; (ii) cultivated crops or fragile vegetation that should not be stepped on by searchers; (iii) standing or running water below the turbines; (iv) inadequate personnel to check the ideal full search area around each turbine; (v) bats or (especially) birds killed by turbines may be blown by the wind to outside the search area boundaries; (vi) birds or bats that are crippled by the turbines may fly on, later dying far outside the wind farm search area; (vii) dense vegetation or other ground cover that conceals carcasses; (viii) bored, inattentive, distracted, or fatigued searchers; (ix) varying levels of experience, skill, and visual acuity among searchers; (x) steep or uneven terrain; (xi) differing weather and light conditions; (xii) small, dull-colored, or partially decomposed carcasses are hard to find; and (xiii) removal of edible carcasses by scavenging animals (perhaps even humans).

To help estimate the real mortality of each bird and bat species, in relation to the observed mortality (i.e., the number of carcasses found), the following simplified equation was developed under this study. (More elaborate equations are found in the scientific literature, for example, Strickland et al. 2009; Kunz, Arnett, Cooper et al. 2007; Smallwood 2007; Anderson et al. 1999; and Morrison 2002.)

$$M = O \ x \ A \ x \ S \ x \ R$$

Where:

M = **Real Mortality**, the number of birds or bats (by species) killed by turbines or other equipment (met towers, transmission lines, and so forth) at a wind farm during a specified time period;

O = **Observed Mortality**, the number of bird or bat carcasses found in the vicinity of turbines or other wind farm equipment during a specified time period;

A = **Area Not Searched**, a factor expressed as $A1/A2$, where $A1$ is the total area of ground in which the bodies of birds or bats killed by wind power equipment are presumed to have fallen and $A2$ is the area of ground actually searched for dead birds or bats by monitoring personnel. The Correction Factor A thus accounts for items i-vi in the first paragraph of this appendix (above);

S = **Searcher Efficiency**, a factor expressed as $S1/S2$, where $S1$ is the total number of bird (by size class) or bat carcasses that are actually lying there (within the area and during the time searched for carcasses) and $S2$ is the number of birds (by size class) or bats that are actually found during the search. The Correction Factor S thus accounts for items vii-xii in the first paragraph; and

R = **Scavenger Removal**, a factor expressed as $R1/R2$, where $R1$ is the total number of birds (by size class) or bats killed by wind power equipment that have fallen within the area searched since the last time that same area was searched, and that would have been found by searchers if scavenging animals had not removed them first, and $R2$ is the number of birds (by size class) or bats that were actually found by the searchers. The Correction Factor R accounts for item xiii in the first paragraph.

Correcting for Factor *A* is relatively straightforward, but it should not be forgotten since many portions of the ideal search area around each monitored structure tend not to be searched, as noted in the first paragraph.

Correcting for Factor *S* requires making assumptions about the real value of *S1*, within different types of search terrain. In general, the taller or denser the vegetation covering the search area, the higher will be the value of *S1* and hence *S*. The bodies of large birds (including most raptors) are relatively more difficult to overlook, so the value of Factor *S* for them is reasonably likely to be between 1.0 and 2.0. However, small-bodied dead birds and bats are much easier to overlook — they are small, motionless, and often camouflaged in color. Young *et al.* (2003) reported searcher detection rates of 92 percent, 87 percent, and 59 percent for large-, medium-, and small-sized birds, respectively at Foote Creek Rim in Wyoming. For Altamont Pass, Smallwood (2007) determined that the average searcher detection efficiency (based on "blind" field trials) was 100 percent for large-bodied raptors, 80 percent for large non-raptor birds, 79 percent for medium sized raptors, 78 percent for medium-sized non-raptor birds, 75 percent for small raptors, and 51 percent for small non-raptor birds. Arnett (2005) found that searchers found only 50 percent of the birds or bats at searcher detection trials conducted in West Virginia. Johnson *et al.* (2002) reported only 38.7 percent searcher efficiency for bird carcass detection at Buffalo Ridge, Minnesota. Bats are widely regarded as even easier to overlook than small birds because the bats lack feathers (that often stick out and catch a searcher's attention) and often resemble dead leaves or rocks. For these reasons, best practice in searching for carcasses, especially of bats, involves the use of trained dogs, which can sniff out many carcasses and thereby substantially reduce the value of *S* (and thus the accuracy of *M*). During the World Bank's October 2008 supervision visit to the La Venta II Project in Mexico, the independent entity responsible for bird and bat monitoring (INECOL) was training a puppy to enhance its carcass search efforts.

Correcting for Factor *R* is challenging and depends upon the local diversity, abundance, and behavior of scavenging animals. Very large birds (including the larger raptors, pelicans, storks, cranes, large waterfowl, and so forth) tend to be scavenged where they fall because they are too heavy or cumbersome to be easily carried away. For these large carcasses, Factor *R* is not likely to exceed 1.0 by very much. However, smaller birds and bats tend to be either rapidly consumed whole, or carried away by the scavengers. To better estimate Factor *R* for smaller carcasses, scavenging trials (using recently killed, fresh bird or bat specimens) are recommended; detailed guidelines and protocols are available from USFWS 2010, CEC and DFG 2007, and PGC 2007 (among others). At La Venta II, a small-scale scavenging trial in 2008 put out three dead bat specimens and checked 24 hours later: Two of the bats had been removed by scavengers and the third remained in place, but had been largely consumed by ants and was much less recognizable as a bat. The La Venta II wind farm has an understandably high scavenging rate because of a large number of scavenging mammals, including Gray Fox *Urocyon cinereoargenteus*, Coyote *Canis latrans*, Northern Raccoon *Procyon lotor*, White-nosed Coati *Nasua narica*, and domestic dogs, as well as scavenging birds such as Turkey Vultures *Cathartes aura*, which are occasionally killed themselves by the turbines. Even outside the tropics, scavenger removal rates for bats tend to be high: At the Mountaineer Wind Energy Center in West Virginia, scavengers removed 25 percent of trial bat specimens in the first 9 hours, 35 percent in one day, and 68 percent within three days (Kerns *et al.*

2005). The time interval between searches can greatly influence the value of Factor R. If searches take place on consecutive days, relatively little data loss (due to carcass removal by scavengers) is likely; conversely, if a week or more elapses between searches, a very high proportion of the carcass evidence is likely to be lost. At La Venta II, the bird and bat mortality searchers in 2008 were checking a given turbine one morning every seven days. Since the available evidence points to very rapid scavenger removal of small carcasses (most gone within 24 hours), it might not be unreasonable to presume—on a precautionary basis, pending further evidence—that Factor R there could be as high as 10, implying that up to 90 percent of the otherwise findable small bird and bat carcasses are removed by scavengers before they can be found by searchers.

These Correction Factors have very important implications in assessing the significance of bird and bat mortality at wind farms. For raptors and other large birds that: (i) are relatively easy to find on the ground when dead and (ii) tend to be scavenged where they fell (so that at least some bones and feathers are usually evident), the real mortality M may not be much higher than the observed mortality O. As an example, the estimated real mortality of Golden Eagles and Red-tailed Hawks at the Altamont Pass Wind Resource Area (WRA) is believed to be less than double the observed mortality (Altamont Pass Monitoring Team 2008). However, for small birds and bats, the difference between observed and real mortality can be very large—up to 1-2 orders of magnitude. For example, at La Venta II, 123 dead bats were found around the turbines during most of 2007 and early 2008. In this area, it is not unreasonable to estimate a value of 2.0 for Correction Factor A, Area Not Searched. During the Bank's 2008 supervision visit, considerable areas were off-limits to searchers because they either had impenetrable native thorn forest vegetation, which has endemic species and conservation value so it should not be cleared from an environmental standpoint, or growing sorghum, which the landowners did not want to be trampled. In addition, the outer limits of the optimum search radius were not searched due to insufficient personnel. For Correction Factor S, Searcher Efficiency, a value of 2.5 for bats seems to be a reasonable first approximation (based on the above-mentioned searcher efficiency data from U.S. wind farms). For Correction Factor R, Scavenger Removal, a preliminary value of 10 seems reasonable (see the preceding paragraph). Since $2 \times 2.5 \times 10 = 50$, finding 123 bat carcasses at La Venta II suggests that upwards of 6,000 bats might have been killed by the project's wind turbines during the approximately one-year study period.

The above-mentioned Correction Factors account for most, though not all, sources of discrepancy between the actual and observed mortality of birds and bats at wind farms. For example, small birds may sometimes disintegrate after being directly hit by a rotor blade, so as to be unrecognizable thereafter on the ground. Conversely, it is theoretically possible that some dead birds or bats found at wind farms might have died of natural causes, unrelated to the turbines or other wind power infrastructure. However, such "baseline mortality" is considered by scientists to be an insignificant source of any carcasses found at wind farms, and can appropriately be disregarded when assessing bird or bat mortality (Johnson et al. 2000).

References

Able, K. 1977. "The flight behavior of individual passerine nocturnal migrants: a tracking radar study." *Animal Behaviour* 25: 924-935.

Ahlen I. 2002. "Fladdermoss och faglar dodade av windkraftverk. [Wind turbines and bats-a pilot study]." *Fauna and Flora* 97: 14–22.

———. 2003. *Wind turbines and bats—a pilot study*. Final Report. Dnr 5210P-2002-00473, P-nr P20272-1.

Ahlen, I., L. Bach, H. J. Baagoe, and J. Pettersson. 2007. Ahlen, I., L. Bach, H. J. Baagoe, and J. Pettersson. Swedish Environmental Protection Agency, Stockholm, Sweden 2007. *Bats and offshore wind turbines studied in southern Scandinavia*. Swedish Environmental Protection Agency, Stockholm, Sweden.

Alcock, J. 1987. "Leks and hilltopping in insects." *Journal of Natural History* 21: 319-328.

Alison, R. 2000. "Shocking findings." (6) in *Wildbird* (July).

Altamont Pass Avian Monitoring Team. 2008. *Altamont Pass Wind Resource Area Bird Fatality Study. July.* (ICF J&S 61119.06.) Portland, OR. Prepared for Altamont County Community Development Agency.

American Wind Energy Association and American Bird Conservancy. 2004. "Wind Energy and Birds/Bats Workshop: Understanding and Resolving Bird and Bat Impacts." Washington, DC May 18-19, 2004. Resolve, Inc. Washington DC.

Anderson, R. L., M. Morrison, K. Sinclair, and M. D. Strickland. 1999. *Studying wind energy-bird interactions: a guidance document. Metrics and methods for determining or monitoring potential impacts on birds at existing and proposed wind energy sites.* Prepared for Avian Subcommittee, National Wind Coordinating Committee, Washington, D.C. December 1999. <http://www.nationalwind.org/publications/wildlife/avian99/Avian_booklet.pdf.>

Anderson, R., N. Neumann, W. P. Erickson, M. D. Strickland, M. Bourassa, K. J. Bay, and K. J. Sernka. 2004. *Avian Monitoring and Risk Assessment at the Tehachapi Pass Wind Resource Area, Period of Performance: October 2, 1996 -May 27, 1998*. NREL Report NREL/SR-500-36416, 90 pp. http://www.nrel.gov/wind/pdfs/36416.pdf.

Anderson, R., J. Tom, N. Neuman, W. P. Erickson, M. D. Strickland, M. Bourassa, K. J. Bay, and K. J. Sernka. 2005. *Avian monitoring and risk assessment at the San Gorgonio Wind Resource Area*. Period of Performance: March 3, 1997-August 11, 2000. NREL/SR50038054. National Renewable Energy Laboratory, Golden, Colorado, USA.

APLIC (Avian Power Line Interaction Committee). 1994. "Mitigating bird collisions with power lines: The state of the art in 1994." U.S. Avian Power Line Interaction Committee.www.aplic.org

———. 2006. "Suggested practices for avian protection on power lines: The state of the art in 2006." U.S. Avian Power Line Interaction Committee. www.aplic.org

Arnett, E. B., editor. 2005. *Relationships between bats and wind turbines in Pennsylvania and West Virginia: an assessment of bat mortality search protocols, patterns of mortality, and behavioral interactions with wind turbines.* A final report submitted to the Bats and Wind Energy Cooperative. Bat Conservation International, Austin, Texas, USA.

Arnett, E. B., K. Brown, W. P. Erickson, J. Fiedler, T. H. Henry, G. D. Johnson, J. Kerns, R. R. Kolford, C. P. Nicholson, T. O'Connell, M. Piorkowski, and R. Tankersley, Jr. 2008. "Patterns of mortality of bats at wind energy facilities in North America." *Journal of Wildlife Management* 72: 61–78.

Arnett, E. B., J. P. Hayes, and M. M. P. Huso. 2006. *Patterns of pre-construction bat activity at a proposed wind facility in south-central Pennsylvania.* An annual report submitted to the Bats and Wind Energy Cooperative. Bat Conservation International, Austin, Texas, USA.

Arnett, E. B., M. M. P. Huso, D. S. Reynolds, and M. Schirmacher. 2007. *Patterns of pre-construction bat activity at a proposed wind facility in northwest Massachusetts.* An annual report submitted to the Bats and Wind Energy Cooperative. Bat Conservation International. Austin, Texas, USA

Arnett, E. B., M. M. P. Huso, M. R. Schirmacher, and J. P. Hayes. 2010. "Changing wind turbine cut-in speed reduces bat fatalities at wind facilities." *Frontiers in Ecology and the Environment 2010*; doi:10.1890/100103.

Arnett, E. B, D. B. Inkley, D. H. Johnson, R. P. Larkin, S. Manes, A. M. Manville, J. R. Mason, M. L. Morrison, M. D. Strickland, and R. Thresher. 2007. "Impacts of wind energy facilities on wildlife and wildlife habitat." *Wildlife Society Technical Review* 07-2. The Wildlife Society, Bethesda, Maryland, USA.

Arnett, E. B., M. Schirmacher, M. M. P. Huso, and J. P. Hayes. 2009. *Effectiveness of changing wind turbine cut-in speed to reduce bat mortalities at wind facilities.* An annual report submitted to the Bats and Wind Energy Cooperative. Bat Conservation International. Austin, Texas, USA. Associated Press. "Audobon Society backs controversial wind farm." 2006, <http://www.msnbc.msn.com/id/12066651/from/ET/> (1 Feb 2010).

AVEN (Agencia Valenciana de la Energía). 2009. *Plan Eólico de la Comunidad Valenciana,* July 31, 2001. < http://www.aven.es/informes/eolico.html#>

Avery, M., and T. Clement. 1972. "Bird mortality at four towers in eastern North Dakota: Fall 1972." *Prairie Naturalist* 4: 87–95.

Baerwald, E. F. 2008. "Variation in the activity and mortality of migratory bats at wind energy facilities in southern Alberta: causes and consequences." Thesis, University of Calgary, Calgary, Alberta, Canada.

Baerwald, E. F., J. Edworthy, M. Holder, and R. M. R. Barclay. 2009. "A large-scale mitigation experiment to reduce bat mortalities at wind energy facilities." *Journal of Wildlife Management* 73:1077-1081.

Barclay, R. M. R., E. F. Baerwald, and J. C. Gruver. 2007. "Variation in bat and bird mortalities at wind energy facilities: assessing the effects of rotor size and tower height." *Canadian Journal of Zoology* 85:381–387.

Barclay, R. M. R., and L. M. Harder. 2003. "Life histories of bats: life in the slow lane." Pages 209–253 in T. H. Kunz and M. B. Fenton, editors, *Bat Ecology.* University of Chicago Press, Chicago, Illinois, USA.

Barlow, K. E. and G. Jones. 1997. "Function of pipistrelle social calls: field data and a playback experiment." *Animal Behaviour* 53: 991–999.

BBC News. 2006. "Wind farm 'hits eagle numbers'. Wind farm turbine blades are killing a key population of Europe's largest bird of prey, UK wildlife campaigners warn." BBC News Friday, June 23, 2006. <http://news.bbc.co.uk/2/hi/europe/5108666.stm>

Behr, O., D. Eder, U. Marckmann, H. Mette-Christ, N. Reisinger, V. Runkel, and O. von Helversen. 2007. "Akustisches Monitoring im Rotorbereich von Windenergieanlagen und methodische Probleme beim Nachweis von Flederamus-Schlagopfern — Ergebnisse aus Untersuchungen im mittleren und südlichen Schwarzwald." *Nyctalus* 12(2-3): 115–127.

Bernis, F. 1980. "La Migración de las aves en el Estrecho de Gibraltar." Vol. I: *Aves Planeadoreas*. Univ. Complutense, Madrid.

Bioscan (UK) Ltd. 2001. *Novar Windfarm Ltd Ornithological Monitoring Studies - Breeding bird and birdstrike monitoring 2001 results and 5-year review*. Report to National Wind Power Ltd.

BioRessource Consultants Ojai, California. Poer Final Project Report. *Identifying Electric Distribution poles for priority retrofitting to reduce bird morality*. California Energy Commission. Public Interest Energy Research Program. April 2008.

BirdLife International. 2000. *Threatened Birds of the World*. Cambridge, UK.

Blanchard, B. 2006. "Land seizures provoke growing anger in China." Reuters, July 7, 2006. < http://www.reuters.com>

Bowman, I., and J. Siderius. 1984. "Management guidelines for the protection of heronries in Ontario." Nongame Program, Wildlife Branch, Ontario Ministry of Natural Resources.

Brinkmann, R., H. Schauer-Weisshahn, and F. Bontadin. 2006. "Survey of possible operational impacts on bats by wind facilities in Southern Germany." Administrative District of Freiburg—Department 56 Conservation and Landscape Management. <http://www.buerobrinkmann.de/downloads/Brinkmann_Schauer-Weisshahn_2006.pdf>. Accessed 1 May 2009.

Brown, W. D., and J. Alcock. 1991. "Hilltopping by the red admiral butterfly: mate seeking alongside congeners." *Journal of Research on the Lepidoptera* 29: 1-10.

Brown, W. K., and B. L. Hamilton. 2002. *Draft report: bird and bat interactions with wind turbines Castle River Wind Farm, Alberta*, Report for VisionQuest Windelectric, Inc., Calgary, Alberta, Canada http://www.batcon.org/home/index.asp?idPage=55&idSubPage=31> Accessed 1 September 2007.

———. 2006a. *Bird and bat monitoring at the McBride Lake Wind Farm, Alberta, 2003-2004*. Report for Vision Quest Windelectric, Inc., Calgary, Alberta, Canada <http://www.batcon.org/home/index.asp?idPage=55&idSubPage=31> Accessed 1 September 2007.

———. 2006b. *Monitoring of bird and bat collisions with wind turbines at the Summerview Wind Power Project, Alberta, 2005-2006*. Report for Vision Quest Windelectric, Inc., Calgary, Alberta, Canada. <http://www.batcon.org/home/index.asp?idPage=55&idSubPage=31> Accessed 1 September 2007.

Bruderer, B. 1994. "Nocturnal migration in the Negev (Israel)–a tracking radar study." *Ostrich* 65:204-212.

Bruderer, B., and A. G. Popa-Lisseanu. 2005. "Radar data on wing-beat frequencies and flight speeds of two bat species." *Acta Chiropterológica* 7:73-82.

Canadian Wildlife Service. 2007. "Wind turbines and birds: a guidance document for environmental assessment." April 2007, *Environment Canada*. <http://www.cwsscf. ec.gc.ca/publications/eval/turb/index_e.cfm>

Carter, T. C., M. A. Menzel, and D. A. Saugey. 2003. "Population trends of solitary foliage-roosting bats." Pages 41–47 in T. J. O'Shea and M. A. Bogan, editors. *Monitoring trends in bat populations of the United States and Territories: problems and prospects.* U.S. Geological Survey, Biological Resources Discipline, Information and Technology Report USGS/BRD/ITR-2003-0003, Washington, D.C., USA.

CEC (California Energy Commission). 2008. "A roadmap for PIER research on methods to assess and mitigate impacts of wind energy development on birds and bats in California." Sacramento: California Energy Commission, CEC-500-2008-076. October 2008.

CEC (California Energy Commission) and DFG (Department of Fish and Game). 2007. "California Guidelines for Reducing Impacts to Birds and Bats from Wind Energy Developments." Sacramento: California Energy Commission and Department of Fish and Game. October 2007.

CFE (Comisión Federal de Electricidad). 2009. "Plan de Vigilancia de la fauna (aves y murciélagos) dentro de la Central Eólica La Venta II, Municipio de Juchitán, Oaxaca: Informe Final 2009." México: Comisión Federal de Electricidad.

Cleveland, C. J., M. Betke, P. Federico, J. D. Frank, T. G. Hallman, J. Horn, J. D. Lopez Jr., G. F. McCracken, R. A. Medellin, A. Moreno-Valdez, C. G. Sansone, J. K. Westbrook, and T. H. Kunz. 2006. "The economic value of pest control services provided by the Brazilian free-tailed bats in south-central Texas." *Frontiers in Ecology and the Environment* 4: 238–243.

CMS (Convention on Migratory Species). 2002a. "Electrocution of Migratory Birds. Bonn, Germany: Convention on the Conservation of Migratory Species of Wild Animals, Resolution 7.4" Adopted by the Conference of the Parties at its Seventh Meeting, 18-24 September 2002.

———. 2002b. "Wind Turbines and Migratory Species. Bonn, Germany: Convention on the Conservation of Migratory Species of Wild Animals, Resolution 7.5, Adopted by the Conference of the Parties at its Seventh Meeting," 18-24 September 2002.

Cole, S. G. 2011. "Wind power compensation is not for the birds: An opinion from an environmental economist." *Restoration Ecology* 19(2)1-17.

CECC (Congressional-Executive Commission on China) Virtual Academy. 2009. "Power Plant Construction Continues After Government Suppresses Villager Protests in Shanwei." <http://www.cecc.gov/pages/virtualAcad/index. phpd?showsingle=34606>

Cooper, B. A., and R. H. Day. 2004. *Results of endangered bird and bat surveys at the proposed Kaheawa Pastures Wind Energy Facility on Maui Island, Hawaii, Fall 2004.* Report prepared for Kahaewa Windpower LLC, Makawao, Hawaii, USA, ABR, Environmental Research & Services, Forest Grove, Oregon, and Fairbanks, Alaska, USA.

Crawford, R. L., and W. W. Baker. 1981. "Bats killed at a north Florida television tower: a 25-year record." *Journal of Mammalogy* 62: 651–652.

Cryan, P. M. 2003. "Seasonal distribution of migratory tree bats (*Lasiurus* and *Lasionycteris*) in North America." *Journal of Mammalogy* 84: 579–593.

———. 2008. "Mating behavior as a possible cause of bat fatalities at wind turbines." *Journal of Wildlife Management* 72: 845–849.

Cryan, P. M., and A. C. Brown. 2007. "Migration of bats past a remote island offers clues toward the problem of bat mortalities at wind turbines." *Biological Conservation* 139: 1–11.

Cryan, P. M., and J. P. Veilleux. 2007. "Migration and use of autumn, winter, and spring roosts by tree bats." M. J. Lacki, A. Kurta, and J. P. Hayes, editors. *Conservation and Management of Bats in Forests*. John Hopkins University Press, Baltimore, Maryland, USA. 153–175

Damborg, S. 1998. *Public Attitudes Towards Wind Power*. Danish Wind Industry Association.

Danish Wind Industry Association. 2008. "Wind Turbine Placement with Regard to Sound Impacts." <http://www.windpower.org>

———. "Shadow Casting from Wind Turbines." 2003. <http://guidedtour.windpower.org/en/tour/env/shadow/index.htm> (7 Jan 2010).

Davis, W. H., and H. B. Hitchcock. 1965. "Biology and migration of the bat, Myotis lucifugus, in New England." *Journal of Mammalogy* 46: 296–313.

De la Torre, A., P. Fajnzylber, and J. Nash. 2009. *Low Carbon, High Growth: Latin American Responses to Climate Change*. Washington: World Bank.

De La Zerda, S., and L. Rosselli. 2003. "Mitigación de colisión de aves contra líneas de transmisión eléctrica con marcaje del cable de guardia (Mitigation of collisions of birds with high-tension electric power lines by marking the ground wire)." *Ornitologia Colombiana* 1(2003): 42–46.

De Lucas, M. G. Janss, and M. Ferrer (eds). 2007. *Birds and wind farms: Risk assessment and mitigation*. Madrid, Quercus: 275.

De Lucas, M. G., F. E. Janss, and M. Ferrer. Department of Applied Biology, "The effects of a wind farm on birds in a migration point: the Strait of Gibraltar." 10 July 2002.

Desholm, M. A., D. Fox, P. D. L. Beasley, and J. Kahlert. 2006. "Remote techniques for counting and estimating the number of bird-wind turbine collisions at sea: a review." *Ibis* 148: 76-89.

"Development on Buffalo Ridge, Minnesota." *Wildlife Society Bulletin* 30: 870887.

Dierschke, V., S. Garthe, and B. Mendel. 2006. "Possible conflicts between offshore wind farms and seabirds in the German sectors of the North Sea and Baltic Sea." *Offshore Wind Energy: Research on Environmental Impacts*. J. Koller, J. Koppel and W. Peters, (eds.). Springer-Verlag Berlin: 371.

Dirksen, S., J. van der Winden, and A. L. Spaans. 1998. "Nocturnal collision risks of birds with wind turbines in tidal and semi-offshore areas." In Ratto, C.F. & Solari, G. (eds.) *Wind Energy and Landscape Balkema, Rotterdam*, ISBN 90 5410 9130, 99-108.

DOE (Department of Energy). 2008. *20% wind energy by 2030: increasing wind energy's contribution to U.S. electricity supply*. U.S. Department of Energy, DOE/GO-102008-256, Washington, D.C.

Dulas Engineering Ltd. 1995. "The Mynydd y Cemmaes windfarm impact study." Vol. IID *Ecological Impact -Final report*. ETSU report: W/13/00300/REP/2D.

Dürr T., 2003. "Kollision von Fledermäuse und Vögel durch Windkraftanlagen." Daten aus Archiv der Staatlichen Vogelschutzwarte Brandenburgs, Buckow.

Dürr, T., and L. Bach. 2004. "Bat deaths and wind turbines——a review of current knowledge, and of the information available in the database for Germany." *Bremer Beiträge für Naturkunde und Naturschutz* 7:253–264 [in German].

EAS (Ecological Advisory Service). 1997. *Ovenden Moor Ornithological Monitoring.* Report to Yorkshire Windpower. Keighley: Ecological Advisory Service.

Eastwood, E. 1967. *Radar ornithology.* Methuen, London, United Kingdom. 290 pp.

EC (European Commission). 2010. *Guidance document: Wind energy developments and Natura 2000.* Brussels: European Commission, 116 pp.

Eckert, H. G. 1982. "Ecological aspects of bat activity rhythms." T. H. Kunz, editor. *Ecology of bats.* Plenum Press, New York, New York, USA. 201–242.

EEA (European Environment Agency). 2009. *Europe's onshore and offshore wind energy potential: An assessment of environmental and economic constraints.* Copenhagen: European Environment Agency, 85 pp.

Environment Canada Canadian Wildlife Service. *Recommended Protocols for Monitoring Impacts of Wind Turbines on Birds.* February 19, 2007

Erickson, W. P., B. Gritski, and K. Kronner. 2003. *Nine Canyon Wind Power Project avian and bat monitoring annual report.* Technical report submitted to energy Northwest and the Nine Canyon Technical Advisory Committee. Western Ecosystems Technology, Inc., Cheyenne, Wyoming, USA.

Erickson, W. P., J. Jeffrey, K. Kronner, and K. Bay 2003. *Stateline Wind Project wildlife monitoring annual report, results for the period July 2001—December 2002.* Technical report submitted to FPL energy, the Oregon Office of Energy, and the Stateline Technical Advisory Committee. Western Ecosystems Technology, Inc., Cheyenne, Wyoming.

Erickson W. P., G. D. Johnson, and D. P. Young Jr. 2005. *A Summary and Comparison of Bird Mortality from Anthropogenic Causes with an Emphasis on Collisions.* USDA Forest Service Gen. Tech. Rep. PSW-GTR-191.

Erickson, W. P., G. D. Johnson, D. Strickland, D. P. Young Jr., K. J. Sernka, and R. E. Good. 2001. *Avian collisions with wind turbines: A summary of existing studies and comparisons to other sources of avian collision mortality in the United States.* National Wind Coordinating Committee, Washington, D.C., USA <http://www.westinc.com/wind_reportreports.php> Accessed 1 September 2007.

Erickson, W. P., G. D. Johnson, D. Young, D. Strickland, R. Good, M. Bourassa, K. Bay and K. Sernka. 2002. *Synthesis and comparison of baseline avian and bat use, raptor nesting and mortality information from proposed and existing wind developments.* Bonneville Power Administration, Portland, Oregon, USA.

Erickson,W. P., and M. D. Strickland. 2007. "Interim Summary on the Effectiveness of the Winter Period Turbine Shutdown in the Altamont Pass Wind Resource Area." Unpublished. Draft Report. April 8, 2007.

Erickson, J. L., and S. D. West. 2002. "The influence of regional climate and nightly weather conditions on activity patterns of insectivorous bats." *Acta Chiropterologica* 4: 17–24

Erp, F. 1997. "Siting Processes for Wind Energy Projects in Germany." Eindhoven University of Technology. In Damborg, S., *Public Attitudes Towards Wind Power.* Danish Wind Industry Association.

European Commission. 2010. "Wind energy developments and Natura 2000." http://ec. europa.eu/environment/nature/natura2000/management/docs/Wind_farms.pdf.

Evans, W. R. 2000. "Applications of acoustic bird monitoring for the wind power industry." In LGL Environmental Research Associates, editors. *Proceedings of National Avian Wind Power Planning Meeting*, 27–29 May 1998, San Diego, California, USA. LGL Environmental Research, King City, Ontario, Canada: 141-152

Everaert, J., K. Devos, and E. Kuijken. 2002. *Windturbines en vogels in Vlaanderen: Voorlopigeonderzoeksresultaten en buitenlandse bevindingen. [Wind turbines and birds in Flanders (Belgium): Preliminary study results in a European context].* Instituut voor Natuurbehoud, Report R.2002.03, Brussels, B: 76.

Everaert. J., and E. Stienen. 2006. "Impact of wind turbines on birds in Zeebrugge (Belgium): Significant effect on breeding tern colony due to collisions." *Biodiversity and Conservation.* DOI 10.1007/s10531-006-9082-1.

Exo, K. M., O. Huppop, and S. Garthe. 2003. "Birds and offshore wind facilities: a hot topic in marine ecology." *Wader Study Group Bulletin* 100: 50–53.

Farnsworth, Q. 2005. "Flight calls and their value for future ornithological studies and conservation research. " *The Auk* 122: 733–746.

Fiedler, J. K. 2004. "Assessment of bat mortality and activity at Buffalo Mountain Windfarm, eastern Tennessee." Thesis, University of Tennessee, Knoxville, Tennessee, USA.

Fiedler, J. K., T. H. Henry, C. P. Nicholson, and R. D. Tankersley. 2007. "Results of bat and bird mortality monitoring at the expanded Buffalo Mountain windfarm, 2005." *Tennessee Valley Authority.* Knoxville, USA http://www.batcon.org/home/index.asp?idPage=55&idSubPage=31> Accessed 1 September 2007.

Finlayson, C. 1992. *Birds of the Strait of Gibraltar.* Poyser, London: 534.

Fisher, Elizabeth, Judith Jones & Rene von Schomberg (eds). 2006. *Implementing the Precautionary Principle: Perspectives and Prospects.* Cheltenham, UK and Northampton, MA, USA: Edward Elgar.

Fleming, T. H., and P. Eby. 2003. "Ecology of bat migration." T. H. Kunz and M. B. Fenton, editors. *Bat Ecology.* University of Chicago Press, Chicago, Illinois, USA. 156–208.

French, H. 2006. "China Covers Up Violent Suppression of Village Protest." *The New York Times,* June 27, 2006. In EastSouthWestNorth: "The Shanwei (Dongzhou) Incident." <http://zonaeuropa.com/20051209_1.htm>

— — —. 2005. "20 Reported Killed as Chinese Unrest Escalates." *The New York Times,* December 9, 2005. In EastSouthWestNorth: "The Shanwei (Dongzhou) Incident." <http://zonaeuropa.com/20051209_1.htm>

Garthe, S., and O. Huppop. 2004. "Scaling possible adverse effects of marine wind farms on seabirds: developing and applying a vulnerability index." *Journal of Applied Ecology* 41:724–734.

Gauthreaux, S. A. 1995. "Suggested practices for monitoring bird populations, movements and mortality in wind Resource Areas." "Proc. National Avian wind power planning meeting," Denver, Colorado, 20–21 July 1994. LGL Ltd., Environmental Research Associates, King City, Ontario.

Gauthreaux, S. A., Jr., and C. G. Belser. 2003. "Radar ornithology and biological conservation." *The Auk* 120: 266–277.

Gauthreaux, S. A., Jr., and J. W. Livingston. 2006. "Monitoring bird migration with a fixed beam radar and a thermal imaging cameras." *Journal of Field Ornithology* 77: 319–328.

Gipe, P. 1995. "Wind Energy Comes of Age," New York. In Damborg, S., *Public Attitudes Towards Wind Power*. Danish Wind Industry Association.

Goodland, R., C. Watson, and G. Ledec. 1984. *Environmental management in tropical agriculture*. Boulder, Colorado: Westview Press, 237p.

Gray, M. 2008. "Tribal Winds Blowing Strong." *Cultural Survival Quarterly* 32(2): 3538.

Griffin, D. R. 1958. *Listening in the dark*. Yale University Press, New Haven, Connecticut, USA.

Griffin, D. R., J. J. G. McCue, and A. D. Grinnell. 1963. "The resistance of bats to jamming." *Journal of Experimental Zoology* 152: 229–250.

Grindal, S. D. 1996. "Habitat use by bats in fragmented forests." R. M. R. Barclay and R. M. Brigham, editors. *Bats and Forests Symposium*. October 19–21, 1995, Victoria, British Columbia, Canada. 260–272.

Grindal, S. D., and R. M. Brigham. 1998. "Short-term effects of small-scale habitat disturbance on activity by insectivorous bats." *Journal of Wildlife Management* 62: 996–1003.

Guillemette, J., J. K. Larsen, and I. Causager. 1998. *Impact assessment of an off-shore wind park on sea ducks*. National Environmental Research institute, Denmark. NERI Technical Report No. 227: pp. 61.

Guillemette, M., Larsen, J. K., and I. Clausager. 1999. *Assessing the impact of the Tunø Knob wind park on sea ducks: the influence of food resources*. NERI Technical Report No. 263. 21pp.

Hall, L. S., and G. C. Richards. 1972. "Notes on Tadarida australis (Chiroptera: Molossidae)." *Australian Mammalogy* 1: 46.

Harley, M., et al. 2001. "Wind farm development and nature conservation: A guidance document for nature conservation organisations and developers when consulting over wind farm proposals in England." Peterborough, UK: English Nature. http://www.bwea.com/pdf/wfd.pdf.

Harmata, A. R., K. M. Podruzny, J. R. Zelenak, and M. L. Morrison. 1999. "Using marine surveillance radar to study bird movements and impact assessment." *Wildlife Society Bulletin* 27: 44–52.

Harness, R., and S. Gombobaatar. 2008. "Raptor electrocutions in the Mongolia steppe." *Winging It: Newsletter of the American Birding Association* 20(6): 1–6.

Hayes, J. P. 2003. "Habitat ecology and conservation of bats in western coniferous forests." C. Zabel and R. G. Anthony, editors. *Mammal community dynamics: management and conservation in the coniferous forests of western North America*. Cambridge University Press, United Kingdom. 81–119.

Hayes, J. P., and S. C. Loeb. 2007. "The influences of forest management on bats in North America." M. J. Lacki, A. Kurta, and J. P. Hayes, editors. *Conservation and management of bats in forests*. John Hopkins University Press, Baltimore, Maryland, USA. 207–235.

Heath, M. F., and M. I. Evans (eds.) 2000. "Important Bird Areas in Europe. Volume I: North Europe." *BirdLife International*. BirdLife Conservation Series No. 8.

Hodos, W. 2003. *Minimization of motion smear: reducing avian collisions with wind turbines.* Period of Performance: July 12, 1999 to August 31, 2002. NREL/SR-500 33249.

Hodos, W., A. Potocki, T. Storm and M. Gaffney. 2001. "Reduction of motion smear to reduce avian collisions with wind turbines." "Proceedings of the National Avian-wind Power Planning Meeting IV," Carmel, Ca, May 16–17, 2000. Prepared for the Avian Subcommittee of the National Wind Coordinating Committee, by RESOLVE, Inc. Washington D.C., 179 pp.

Hoover, S. L., and M. L. Morrison. 2005. "Behavior of red-tailed hawks in a wind turbine development." *Journal of Wildlife Management* 69(1): 150–159.

Horn, J. W, E. B. Arnett, and T. H. Kunz. 2008a. "Behavioral responses of bats to operating wind turbines." *Journal of Wildlife Management* 72: 123–132.

———. 2008b. *Behavioral responses of bats to testing the effectiveness of an experimental acoustic bat deterrent at the Maple Ridge wind farm.* An annual report submitted to the Bats and Wind Energy Cooperative. Bat Conservation International. Austin, Texas, USA.

———. 2008c. "Analysing NEXRAD Doppler radar images to assess nightly dispersal patterns and population trends in Brazilian free tailed bats (Tadarida brasiliensis)." *Integrative and Comparative Biology,* 48: 1–11, doi:10.1093/icb/icn051.

Hötker, H., K. M. Thomsen, and H. Jeromin. 2006. *Impacts on biodiversity of exploitation of renewable energy resources: the example of birds and bats.* Michael-Otto-Institut im NABU, Bergenhusen.

Hötker, H., K. M. Thomsen, and H. Köster. 2004. *Auswirkungen regenerativer Energiegewinnung auf die biologische Vielfalt am Beispiel der Vögel und der Fledermäuse—Fakten, Wissenslücken, Anforderungen an die Forschung, ornitologische Kriterien zum Ausbau von regenerativen Energiegewinnungsformen.* Report requested by Bundesamt für Naturschutz, Förd Nr. Z1.3-684 11-5/03. NABU, Germany.

Howe, R. W., W. Evans, and A. T. Wolf. 2002. *Effects of wind turbines on birds and bats in northeastern Wisconsin.* Wisconsin Public Service Corporation, Green Bay, USA.

Howell, J. A., and J. Noone. 1992. *Examination of avian use and mortality at a U.S. Windpower wind energy development site,* Solano County, California. Final Report to Solano County Department of Environmental Management, Fairfield, California, USA.

Hristov, N. I., and W. E. Conner. 2005. "Sound strategy: acoustic aposematism in the bat-tiger moth arms race." *Naturwissenschaften* 92: 164–169.

Hristov, N. I., M. Betke, and T. H. Kunz. 2008 "Applications of thermal infrared imaging for research in aeroecology." *Integrative and Comparative Biology,* 48: 50–59, doi: 10.1093/icb/icn053. <http://ec.europa.eu/environment/nature/natura2000/management/docs/Wind_farms.pdf>

Humphrey, S. R., and J. B. Cope. 1976. "Population ecology of the little brown bat, Myotis lucifugus, in Indiana and north-central Kentucky." *American Society of Mammalogists Special Publication No. 4.*

Hunt, W. G. 1995. *A pilot golden eagle population project in the Altamont Pass wind Resource Area, California,* prepared by the Predatory Bird Research Group, University of California, Santa Cruz, for the National Renewable Energy Laboratory, Golden, Colorado.

Iberdrola. 2008. "Avian and Bat Protection Plan." *Iberdrola Renewables.*

IEC (International Electrotechnical Commission). 2002. *Wind Turbine Generator Systems-Part 11: Acoustic noise measurement techniques.* International Standard 61400-11, 2nd ed.

IFC (International Finance Corporation). 2007. *Stakeholder Engagement: A Good Practice Handbook for Companies Doing Business in Emerging Markets.* Washington, DC.

INECOL (Instituto Nacional de Ecología). 2009. *Plan de Vigilancia de la Fauna (Aves y Murciélagos) dentro de la Central Eoloeléctrica La Venta II, Municipio de Juchitán, Oaxaca: Informe Final 2009.* Veracruz, México: Instituto Nacional de Ecología.

IPCC (Intergovernmental Panel on Climate Change). 2007. *Synthesis Report: An Assessment of the Intergovernmental Panel on Climate Change.* Geneva, Switzerland.

Jain, A., P. Kerlinger, R. Curry, and L. Slobodnik. 2007. *Annual report for the Maple Ridge wind power project post-construction bird and bat mortality study—2006.* Annual report prepared for PPM Energy and Horizon Energy, Curry and Kerlinger LLC, Cape May Point, New Jersey, USA

Janss, G. 1998. "Bird behavior in and near a wind farm at Tarifa, Spain: management considerations." NWCC National Avian - Wind Power Planning Meeting III, 110–114.

———. 2000. "Bird Behaviour in and Near a Wind Farm at Tarifa, Spain: Management Considerations." In *Proceedings of the National Avian-Wind Power Planning Meeting III*, pp. 110–114. <http//www.nrel.gov>.

Jenkins, A., and J. Smallie. 2009. "Terminal velocity: End of the line for Ludwig's Bustard?" *Africa Birds and Birding* 14(2): 35–39.

Johnson, G. D. 2005. "A review of bat mortality at wind-energy developments in the United States." *Bat Research News* 46: 45–49.

Johnson, G. D., W. P. Erickson, M. D. Strickland, M. F. Shepherd and D. A. Shepherd. 2000. *Avian monitoring studies at the Buffalo Ridge Wind Resource Area, Minnesota: results of a four-year study.* Technical report prepared for Northern States Power Co., Minneapolis, Minnesota. Western Ecosystems Technology, Inc., Cheyenne, Wyoming, USA.

Johnson, G. D., W. P. Erickson, M.D. Strickland, M. F. Shepherd, and S. A. Sarappo. 2002. "Collision mortality of local and migrant birds at a large-scale wind-power development on Buffalo Ridge, Minnesota." Wildlife Society Bulletin 30:870-887.

———. 2003. "Mortality of bats at a large-scale wind power development at Buffalo Ridge, Minnesota." *American Midland Naturalist* 150: 332–342.

Johnson, G. D., M. K. Perlik, W. P. Erickson, and M. D. Strickland. 2004. "Bat activity, composition and collision mortality at a large wind plant in Minnesota." *Wildlife Society Bulletin* 32: 1278–1288.

Kerlinger, P. 1980. "A tracking radar study of bird migration." *Journal of Hawk Migration Association of North America* 2: 34–42.

———. 2001. "Avian issues and potential impacts associated with wind power development of nearshore waters of Long Island, New York." Prepared for Bruce Bailey, AWS Scientific.

———. 2003. Avian Risk Assessment for the East Haven Wind farm, East demonstration Project, Essex County, Vermont. Report prepared for East Haven Wind farm, Matthew Rubin, Project Manager. http://easthavenwindfarm.com/filing.hml

Kerlinger, P., and J. Dowdell. 2003. "Breeding bird survey for the Flat rock wind power project, Lewis County, New York." Prepared for Atlantic Renewable Energy Corporation. <www.Flatrockwind.com/documents/flatrocknestRpt-9-03.pdf>

Kerns, J., and P. Kerlinger. 2004. *A study of bird and bat collision mortalities at the Mountaineer Wind Energy Center, Tucker County, West Virginia.* Annual Report for 2003. Curry and Kerlinger, L. L. C., McLean, Virginia, USA. <http://www.batcon.org/home/index.asp?idPage=55&idSubPage=31.> Accessed 1 September 2007.

Kerns, J., W. P. Erickson, and E. B. Arnett. 2005. "Bat and bird mortality at wind energy facilities in Pennsylvania and West Virginia." Pages 24–95 in E. B. Arnett, editor. *Relationships between bats and wind turbines in Pennsylvania and West Virginia: an assessment of bat mortality search protocols, patterns of mortality, and behavioral interactions with wind turbines. A final report submitted to the Bats and Wind Energy Cooperative.* Bat Conservation International, Austin, Texas, USA. <http://www.batcon.org/home/index.asp?idPage=55&idSubPage=31.> Accessed 1 September 2007.

King, J. 2009. "Hawaii in Early Stages of Energy Revolution." CNN, December 18, 2009. <www.cnn.com/2009/POLITICS/12/18/king.sotu.hawaii/index.html>

Kingsley, A., and B. Whittam. 2007. "Wind turbines and birds: a background review for environmental assessment." Canadian Wildlife Service, Environment Canada. Gatineau, Quebec. 131 pp. http://www.nationalwind.org/workgroups/wildlife/canada/Wint_Turbines)and_Birds_Background_Review_2007.pdf.

Koop, B. 1997. "Vogelzug und Windernergieplanung. Beispiele fur Auswirkungen aus dem Kreis Plon (Schleswig-Holstein)." *Naturschutz und Landschaftsplanung* 29(7): 202–207.

Krohn, S., and S. Damborg. 1999. "On Public Attitudes Towards Wind Power." *Renewable Energy* 16(1–4): 954–960.

Kunz, T. H., E. B. Arnett, B. M. Cooper, W. P. Erickson, R. P. Larkin, T. Mabee, M. L. Morrison, M. D. Strickland, and J. M. Szewczak. 2007. "Methods and metrics for studying impacts of wind energy development on nocturnal birds and bats." *Journal of Wildlife Management* 71: 2449–2486.

Kunz, T. H., E. B. Arnett, W. P. Erickson, A. R. Hoar, G. D. Johnson, R. P. Larkin, M. D. Strickland, R. W. Thresher, and M. D. Tuttle. 2007. "Ecological impacts of wind energy development on bats: questions, research needs, and hypotheses." *Frontiers in Ecology and the Environment* 5: 315–324.

Kunz, T. H., S. A. Gauthreaux, Jr., N. I. Hristov, J. W. Horn, G. Jones, E. K. V. Kalko, R. P. Larkin, F. McCraken, S. W. Swartz, R. B. Srygley, R. Dudley, J. K. Westbrook, and M. Wikelski. 2008. "Aeroecology: probing and modelling the aerosphere." *Integrative and Comparative Biology* 48: 1–11, doi: 10.1093/icb/icn037.

Kuran, T. 2004. "Cultural Obstacles to Economic Development: Often Overstated, Usually Transitory." In Rao, V. and M. Walton (eds.), *Culture and Public Action.* Stanford Social Sciences, Stanford, CA, USA.

Lacki, M. J., J. P. Hayes, and A. Kurta, editors. 2007. *Bats in forests: conservation and management.* John Hopkins University Press, Baltimore, Maryland, USA.

Ladenburg, J. 2009. "Visual Impact Assessment of Offshore Wind Farms and Prior Experience." *Applied Energy* 86(3): 380–387.

Ladenburg, J., and A. Dubgaard. 2007. "Willingness to Pay for Reduced Visual Disamenities from Off-Shore Wind Farms in Denmark." *Energy Policy* 35(8): 4059–4071.

Ladenburg, J., A. Dubgaard, L. Martinsen, and J. Tranberg. 2005. *Economic Valuation of the Visual Externalities of Offshore Wind Farms,* Report from the Food and Resource Economic Institute, The Royal Veterinary and Agricultural University, Copenhagen, Report no. 179.

Langston, R., H. W. and J. D. Pullan. 2003. "Windfarms and birds: An analysis of the effects of windfarms on birds, and guidance on environmental assessment criteria and site selection issues." Report written for Birdlife on behalf of the Bern Convention. Convention on the conservation of European wildlife and natural habitats, Standing committee 22nd meeting.

Lanyon, W. E. 1994. Western Meadowlark. A. Poole and F. Gill, editors. *The Birds of North America. No. 104.* The American Ornithologists' Union and The Academy of Natural Sciences, Philadelphia, Pennsylvania, USA.

Larkin, R. P. 1991. "Flight speeds observed with radar, a correction: slow "birds" are insects." *Behavioral Ecology and Sociobiology* 29: 221–224. Cardiff.

———. 2005. "Radar techniques for wildlife biology." C. E. Braun, editor. *Techniques for wildlife investigations and management.* The Wildlife Society, Bethesda, Maryland, USA: 448–464.

———. 2006. "Migrating bats interacting with wind turbines: What birds can tell us." *Bat Research News* 47: 23–32.

Larsson, A. K. 1994. "The Environmental Impact from an Offshore Plant." *Wind Engineering* 18: 213–219.

Lebbin, D., M. Parr, and G. Fenwick. 2010. *The American Bird Conservancy Guide to Bird Conservation.* Lynx Edicions, Barcelona and the University of Chicago Press, Chicago and London.

Ledec, G. "Conservación de la biodiversidad en los proyectos viales: una visión general." Paper presented at "VII Encuentro Latinoamericano de Unidades Ambientales del Sector Transporte (SLUAT)." Lima, Perú. 2005.

Ledec, G., and P. Posas. 2003. "Biodiversity Conservation in Road Projects: Lessons from World Bank Experience in Latin America (198–202) in *Transportation Research Record* 1819 (1).

Lekuona, J. M. 2001. *Uso del espacio por la avifauna y control de la mortalidad de aves y murciélagos en los parques eólicos de Navarra durante un ciclo anual.* Gobierno de Navarra, Spain. http://www.iberica2000.org/textos/LEKUONA_REPORT.pdf.

Lekuona, J. M., and C. Ursua. 2007. "Avian mortality in wind power plants of Navarra (northern Spain)" in M. de Lucas, G.F.E. Janss, and M. Ferrer, eds., *Birds and Wind Farms: Risk Assessment and Mitigation.* Madrid: Quercus.

Liechti, F., B. Bruderer, and H. Paproth. 1995. "Quantification of nocturnal bird migration by moonwatching: comparison with radar and infrared observations." *Journal of Field Ornithology* 66: 457–468.

Longcore T., C. Rich, and S. A. Gauthreaux, Jr. 2005. *Scientific Basis to Establish Policy Regulating Communications Towers to Protect Migratory Birds: Response to Avatar Environmental, LLC, Report Regarding Migratory Bird Collisions With Communication Towers,* WT Docket No. 03-187, Federal Communications Commission Notice of Inquiry. American Bird Conservancy, Defenders of Wildlife, Forest Conservation Council, The Humane Society of the United States. February 14.

Lowther, S. 2000. "The European perspective: some lessons from case studies." In *Proceedings of National Avian Paper presented at the Wind Power Planning Meeting III*, San Diego, California May 1998. Prepared for the Avian Subcommittee of the National Wind Coordinating Committee by LGL Ltd., King City, Ontario. 202 pp.

Mabee, T. J., and B. A. Cooper. 2004. "Nocturnal bird migration in northeastern Oregon and southeastern Washington." *Northwestern Naturalist* 85: 39–47.

Mabee, T. J., J. H. Plissner, B. A. Cooper, and D. P. Young. 2006. "Nocturnal bird migration over an Appalachian ridge at a proposed wind power project." *Wildlife Society Bulletin* 34: 682–690.

Mackey, R. L., and R. M. R. Barclay. 1989. "The influence of physical clutter and noise on the activity of bats over water." *Canadian Journal of Zoology* 67: 1167–1170.

Manes, R., S. Harmon, B. Obermeyer and R. Applegate. 2002. "Wind energy and wildlife: an attempt at pragmatism." Wildlife Management Institute <http://www.wildlifemanagementinstitute.org/pages/windpower.html\>.

Manwell, J. F., J. G. McGowan, and A. L. Rogers. 2002. *Wind Energy Explained: Theory, Design and Application.* John Wiley & Sons, Inc., Hoboken, N.J. USA.

Marti, R. 1995. *Bird/wind turbine investigations in southern Spain.* Prepared for the National Avian—Wind Power Planning Meeting, Denver, Colorado, July 1994. "RESOLVE Inc.," Washington D.C. and LGL Ltd., King City, Ontario, 145 pp.

Marti, R. and L. Barrios. 1995. *Effects of wind turbine power plants on the avifauna in the Campo de Gibraltar Region—Summary of final report.* Prepared for the Environment Agency of the Regional Government of Andalusia and the Spanish Ornithological Society (SEO/Birdlife): 20.

McIsaac, H. P. 2001. *Raptor acuity and wind turbine blade conspicuity. In Proceedings of the National Avian—Wind Power Planning Meeting IV*, Carmel, CA May 16-17, 2000. Prepared for the Avian Subcommittee of the National Wind Coordinating Committee, by RESOLVE, Inc., Washington D.C. 179 pp. <www.nationalwind.org/pubs/avian00/avian_proceedings_2000.pdf>.

Meek, E. R., J. B. Ribbands, W. G. Christer, P. R. Davy, and I. Higginson. 1993. "The effects of aero-generators on moorland bird populations in the Orkney Islands, Scotland." *Bird Study* 40:140-143.

Merck, T., and H. Nordheim. 1999. "Nature conservation problems arising from the use of offshore wind energy. Actual problems of the marine environment." Lectures at the 9th Scientific Symposium 26–27 May 1999 in Hamburg. Supplement. Hamburg [Dtsch. Hydrogr. Z. (Suppl.)].

Miller, A. 2008. "Patterns of avian and bat mortality at a utility-scaled wind farm on the southern high plains." Thesis, Texas Tech University, Lubbock, Texas, USA.

Molvar, E. M. 2008. "Wind power in Wyoming: Doing it smart from the start." Laramie, WY: Biodiversity Conservation Alliance, 55p. Voice for the Wild. <www.voiceforthewild.org/blm/pubs/WindPowerReportReport.pdf>

Morrison, M. L. 1996. *Protocols for evaluation of existing wind developments and determination of bird mortality. In Proceedings of National Avian - Wind Power Planning Meeting II*, Palm Springs, California, September. Prepared for the Avian Subcommittee of the National Wind Coordinating Committee by RESOLVE Inc., Washington, D.C., and LGL Ltd., King City, Ontario. 152 pp.

———. 2002a. *Avian risk fatality protocol*. National Renewable Energy Laboratory, NREL/SR500-24997. Golden, Colorado, USA.

———. 2002b. *Searcher bias and scavenging rates in bird-wind energy studies*. National Renewable Energy Laboratory, NREL/SR-500-30876. Golden, Colorado, USA.

Morrison, M. L., W. M. Block, M. D. Strickland, and W. L. Kendell. 2001. *Wildlife study design*. Springer Verlag, New York, New York.

Mossop, D. H. 1997. "Five Years of Monitoring Bird Strike Potential at a Mountain-Top Wind Turbine, Yukon Territory." Presented at the Annual Canadian Wind Energy Conference & Exhibition. Yukon Energy Corporation, Whitehorse, Yukon: 197–211.

———. 1998. *Five years of monitoring bird strike potential at a mountain-top wind turbine, Yukon Territory*. Prepared for CANMET Energy Technology Centre, Natural Resources Canada.

Mumford, R. E., and J. O. Whitaker, Jr. 1982. *Mammals of Indiana*. Indian University Press, Bloomington, IN, USA.

Musters, C. J. M., M. A. W. Noordervliet, and W. J. ter Keurs. 1996. "Bird casualties caused by a wind energy project in an estuary." *Bird Study* 43: 124–126.

NABU. 2004. "Auswirkungen regenerativer Energiegewinnung auf die biologische Vielfalt am Beispiel der Vögel und der Fledermäuse—Fakten, Wissenslücken, Anforderungen an die Forschung, ornithologische kriterien zum Ausbau von regenerativern Energiegewinnungsformen." *Bundesamt für Naturschutz; Förd.* Nr. Z1.3 684 11-5/03.

NEFMC (New England Fishery Management Council). 2008. Letter from J. Pappalardo to R. E. Cluck of the U.S. Department of the Interior and R. J. DeSista of the U.S. Army Corps of Engineers, 21 April 2008. <http://www.nefmc.org/habitat/080403%20Cluck%20re%20Cape%20Wind%20comments_final.pdf>.

NRC (National Research Council). T.H. Kunz, et al. 2007. *Environmental Impacts of Wind-Energy Projects*. The National Academies Press, Washington, D.C., 376 pp.

Norwegian Ornithological Society 2001. *Plans to build a wind farm on Smola Archipelago (Norway)*. Submitted to the Convention on the Conservation of European Wildlife and Natural Habitats. Strasbourg, 27 September 2001. <http://www.semantise.com/~lewiswindfarms/FOV1-00021BAE/FOV1-00021D1F/2001:11:26%20Berne-ConventionSmola.pdf?FCItemID=S000C7AA4>

NWCC (National Wind Coordinating Cooperative). 2007. *Assessing impacts of wind-energy development on nocturnally active birds and bats: a Guidance document*. <http://www.nationalwind.org/pdf/Nocturnal_MM_Final-JWM.pdf>.

NYSERDA (New York State Energy Research and Development Authority). 2009. "Comparison of reported effects and risk to vertebrate wildlife from six electricity generation types in the New York/New England Region." Report 90-92 March 2009. NYSERDA 9675.

OFA (Ontario Federation of Agriculture). 2011. "30 Suggestions on Wind Power Leases for Farmers." < http://www.ofa.on.ca/index.php?p=77&a=1063>.

Orloff, S., and A. Flannery. 1992. *Wind turbine effects on avian activity, habitat use and mortality in Altamont Pass and Solano County Wind Resource Areas, 1989-1991*. Report to the Planning Departments of Alameda, Contra Costa and Solano counties and the California Energy Commission, Sacramento, California, USA.

Orloff, S. and A. Flannery. 1996. *A continued examination of avian mortality in the Altamont Pass Wind Resource Area*. Final Report to the California Energy Commission by BioSystems Analysis, Inc., Tiburon, California, USA.

Painter, S., B. Little, and S. Lawrence. 1999. *Continuation of bird studies at Blyth Harbour wind farm and the implications for offshore wind farms*. Report by Border Wind Limited to UK Department of Trade & Industry, ETSU W/13/00485/00/00.

Parsons, S., and J. M. Szewczak. 2008. "Detecting, recording, and analyzing the vocalizations of bats." In T. H. Kunz and S. Parsons, editors. *Ecological and behavioral methods for the study of bats*. Second Edition. Johns Hopkins University Press, Baltimore, Maryland, USA.

Patriquin, K. J., and R. M. R. Barclay. 2003. "Foraging by bats in cleared, thinned and unharvested boreal forest." *Journal of Applied Ecology* 40: 646–657.

Pedersen, E. 2004. *Noise Annoyance from Wind Turbines: A Review*. Report No. 5308, Naturvårdsverket, Swedish Environmental Protection Agency.

Pedersen, M. B., and E. Poulsen. 1991. "En 90m/2MW vindmølles indvirkning på fuglelivet. Fugles reaktioner på opførelsen og idriftsættelsen af Tjæreborgmøllen ved Danske Vadehav (Danish with English summary)." Danske Vildtundersøgelser, Hæfte 47, Danmarks Miljøundersøgelser, Afdeling for Flora-og Faunaøkologi, Kalø.

Percival, S. M. 1998. "Birds and wind turbines—managing potential planning issues." Pages 345–350 in S. Powles, editor. *British Wind Energy Association*. Bury St. Edmunds, UK.

———. 2001. *Assessment of the effects of offshore wind farms on birds*. Report ETSU W/13/00565/REP, DTI/Pub URN 01/1434.

———. "Birds and wind farms in Ireland: A review of potential issues and impact assessment." Unpublished draft.

Pettersson, J., and T. Stalin. 2003. *Influence of offshore windmills on migration birds in southeast coast of Sweden*. Report to GE Wind Energy.

PGC (Pennsylvania Game Commission). 2007. *Pennsylvania Game Commission wind energy voluntary agreement*. Pennsylvania Game Commission, Harrisburg, Pennsylvania, USA. http://www.pgc.state.pa.us/pgc/cwp/view.asp?a=483&Q=175703&PM=1&pp=12&n=1

Piorkowski, M. D. 2006. "Breeding bird habitat use and turbine collisions of birds and bats located at a wind farm in Oklahoma mixed-grass prairie." Thesis, Oklahoma State University, Stillwater, USA.

Pruett, C. L., M. A. Patten, and H. Donald. 2009. "It's Not Easy Being Green: Wind energy and a Declining Grassland Bird." *BioScience* 59(3): 257–252.

PWEA. 2008. "Guidelines for Assessment of Wind Farms' Impact on Birds." Szczecin: Polish Wind Energy Association, 26p.

Racey, P. A., and A. C. Entwistle. 2003. "Conservation ecology of bats." T. H. Kunz and M. B. Fenton, editors. *Bat Ecology*. University of Chicago Press, Chicago, Illinois, USA. 680743.

Raffensperger, C., and J. Tickner, eds. 1999. *Protecting Public Health and the Environment: Implementing the Precautionary Principle*. Island Press, Washington, DC.

Ralph, C. J., S. Droege, and J. R. Sauer. 1995. "Managing and monitoring birds using bird point counts; standards and applications." In J. R. Sauer, S. Droege, eds., *Moni-*

toring bird populations by point counts, General Technical Report PSW-GTR-149. Albany, CA Southwest Research Station, Forest Service, U.S. Department of Agriculture.

Ralph, C. J., G. R. Guepel, P. Pyle, T. E. Martin, and D. F. DeSante. 1993. *Handbook of field methods for monitoring land birds*. Gen. Tech. Rep. PSW-GTR-144, Albany, CA, Pacific Southwest Research Station, Forest Service, U.S. Department of Agriculture. Report to Windcluster, Ltd.

Rao, V., and M. Walton, eds. 2004. *Culture and Public Action*. Stanford Social Sciences, Stanford, CA, USA.

Reynolds, D. S. 2006. "Monitoring the potential impact of a wind development site on bats in the northeast." *Journal of Wildlife Management* 70: 1219–1227.

Rodrigues, L., L. Bach, M. J. Dubourg-Savage, J. Goodwin, and C. Harbusch. 2008. "Guidelines for consideration of bats in wind farm projects." *EUROBATS* Vol. 3 (English version). UNEP/EUROBATS Secretariat, Bonn, Germany.

Rodriguez, E., G. Tiscornia, and L. Olivera. 2009. *Diagnóstico de las Aves y Mamíferos Voladores que Habitan en el Entorno de la Sierra de los Caracoles y el Diseño de Un Plan de Monitoreo: Informe Final*. Montevideo, Uruguay: Administración Nacional de Usinas y Transmisiones Eléctricas, 49 p. <www.ute.com.uy>

Russo, D., G. Jones, and A. Migliozzi. 2002. "Habitat selection by the Mediterranean horseshoe bat, Rhinolophus euryale (Chiroptera: Rhinolophidae) in a rural area of southern Italy and implications for conservation." *Biological Conversation* 107: 71–81.

Salm, R. V., and J. R. Clark. 2000. Marine and Coastal Protected Areas: A Guide for Planners and Managers. Gland, Switzerland: International Union for Conservation of Nature and Natural Resourcs (IUCN). 370 pp.

Seiche, K., P. Endl, and M. Lein. 2007. "Fledermäuse und Windenergieanlagen in Sachsen Ergebnisse einer landesweiten Studie." *Nyctalus* (N.F.) 12 (2/3): 170–181.

SEO/BirdLife. 1995. "Effects of wind turbine power plants on the avifauna in the Campo de Gibraltar region. Summary of final report commissioned by the Environmental Agency of the Regional Government of Andalusia." Unpublished, Sociedad Española de Ornitología, Madrid.

Smallwood, K. S. 2006. *Biological effects of repowering a portion of the Altamont Wind Resource Area, California: The Diablo Winds Energy Project*. Unpublished Report. July 27, 2006.

———. "Estimating wind turbine-caused bird mortality." *Journal of Wildlife Management* 71(8): 2781–2791.

Smallwood, K. S., and C. G. Thelander. 2004. "Developing methods to reduce bird mortality in the Altamont Pass Wind Resource Area." Final report to the California Energy Commission, PIER-EA contract No. 500-01-019, Sacramento, California, USA.

———. 2008. "Bird Mortality in the Altamont Pass Wind Resource Area, California." *Journal of Wildlife Management* 72(1): 215–223.

———. 2009. "Avian and bat fatality rates at old-generation and repowered wind turbines in California." *Journal of Wildlife Management*. 73(7): 1062–1071.

Spaans, A., L. v. d. Bergh, S. Dirksen, and J. v. d. Winden. 1998. *Wind turbines and birds: can they co-exist? De Levende Natur*: 115–121.

Spanjer, G. R. 2006. *Responses of the big brown bat, Eptesicus fuscus, to a proposed acoustic deterrent device in a lab setting.* A report submitted to the Bats and Wind Energy Cooperative and the Maryland Department of Natural Resources. Bat Conservation International. Austin, Texas, USA.

Stewart, G. B., A. S. Pullin, and C. F. Coles. 2004. *Effects of wind turbines on bird abundance Review report.* Centre for Evidence-based Conservation, Systematic Review No. 4, University of Birmingham, Edgbaston, Birmingham, United Kingdom.

Still, D., B. Little, and S. Lawrence. 1995. *The Effect of Wind Turbines on the Bird Population at Blyth,* ETSU Report w/13/00394.

Strickland, M. D., E. B. Arnett, W. P. Erickson, D. H. Johnson, G. D. Johnson, M. L. Morrison, J. A. Shaffer, and W. Warren-Hicks. 2009. *Studying wind energy/wildlife interactions: a guidance document.* Prepared for the National Wind Coordinating Collaborative, Washington DC, USA.

Szewczak, J. M., and E. B. Arnett. 2006. *Preliminary field test results of an acoustic deterrent with the potential to reduce bat mortality from wind turbines.* A report submitted to the Bats and Wind Energy Cooperative. Bat Conservation International. Austin, Texas, USA.

———. 2007. *Field test results of a potential acoustic deterrent to reduce bat mortality from wind turbines.* A report submitted to the Bats and Wind Energy Cooperative. Bat Conservation International. Austin, Texas, USA.

Thelander, C. G., and L. Rugge, 2000a. *Avian risk behavior and fatalities at the Altamont Wind Resource Area—March 1998—February 1999.* Prepared by BioResource Consultants for the National Renewable Energy Laboratory, Subcontract No. TAT-8 18209-01, NREL/SR-500-27545. Golden, Colorado.

———. 2000b. "Bird risk behaviors and mortalities at the Altamont Wind Resource Area". Proceedings of the National Avian-Wind Power Planning Meeting III. National Wind Coordinating Committee/RESOLVE. Washington, D.C., USA. <http://www.nationalwind.org/publications/wildlife/avian98/02-Thelander-Rugge-Altamont.pdf>. Accessed 1 September 2007: 5–14.

Timm, R. M. 1989. "Migration and molt patterns of red bats, Lasiurus borealis (Chiroptera: Vespertilionidae), in Illinois." *Bulletin of the Chicago Academy of Sciences* 14: 1–7.

TNC (The Nature Conservancy). 2008. "Expertise contributes to landmark wind-energy decision." *The Nature Conservancy in Virginia Magazine* Spring/Summer 2008: 12.

Transportation Research Record 1819. 2003. Paper NO. LVR8-1154. Eighth International Conference on Low-Volume Roads June 22–25, Reno, Nevada.

Tuck, L., J. Schwartz, and L. Andres. 2009. *Crisis In LAC: Infrastructure Investment and the Potential for Employment Generation.* LCR Crisis Brief. Washington, DC: World Bank Group.

Tulp, I., H. Schekkerman, J. K. Larsen, J. van der Winden, R. J. W. van de Haterd, P. van Horssen, S. Dirksen, and A. L. Spaans. 1999. *Nocturnal flight activity of sea ducks near the wind farm Tunø Knob in the Kattegat.* IBN-DLO Report No. 99.30. www.alterra.nl.

Tyler, S. J. 1995. *Bird strike study at Bryn Titli windfarm, Rhayader.* Report to National Windpower. 2pp.

UN (United Nations). 2007. *Declaration on the Rights of Indigenous Peoples*. GA Res. 61/295, UN GAOR 61st sess., 107th plen. mtg., UN Doc. A/RES/61/295 (13 September 2007).

USFWS (United States Fish and Wildlife Service). 2010. *Wind Turbine Guidelines Advisory Committee Recommendations*. U.S. Fish and Wildlife Service. Washington, DC.

Usgaard, R. E., D. E. Naugle, R. G. Osborn, and K. F. Higgins. 1997. "Effects of wind turbines on nesting raptors at Buffalo Ridge in southwestern Minnesota." *Proceedings of the South Dakota Academy of Sciences* 76: 113–117.

Van den Bergh, L. M. J., A. L. Spaans, and N. D. van Swelm. 2002. "Lijnopstellingen van windturbines geen barrier voor voedselvluchten van meeuwen en sterns in de broedtijd." *Limosa* 75: 25–32.

Van Rooyen. 1998. "Down the line: The Eskom/EWT Strategic Partnership." *Africa Birds and Birding* 3(6): 14–15.

Vaughn, C. R. 1985. "Birds and insects as radar targets: a review." *Proceedings of the Institute of Electrical and Electronics Engineers* 3: 205–227.

Verboom, B., and H. Huitema. 1997. "The importance to linear landscape elements for the pipistrelle Pipistrellus pipstrellus and the serotine bat Eptesicus serotinus." *Landscape Ecology* 12: 117–125.

Verboom, B., and K. Spoelstra. 1999. "Effects of food abundance and wind on the use of tree lines by an insectivorous bat, Pipistrellus pipistrellus." *Canadian Journal of Zoology* 77: 1393–1401.

Watson, D. M. 2003. "The standardized search: an improved way to conduct bird surveys." *Austral Ecology* 28(5): 515.

Winhold, L., A. Kurta, and R. Foster. 2008. "Long-term change in an assemblage of North American bats: are eastern red bats declining?" *Acta Chiropterológica* 10: 359–366.

Winkelman, J. E. 1989. "Birds and the wind park near Urk: collision victims and disturbance of ducks, geese and swans." RIN Rep. 89/15. *Rijkinstituut voor Natuurbeheer*, Arnhem, The Netherlands. English summary.

— — —.1992. "De invloed van de Sep-proefwindcentrale te OOsterbierum (Fr.) op vogels, 1. Aanvaringsslachtoffers (The impact of the Sep Wind Park near Oosterbierum [Fr.], the Netherlands, on birds, 1. Collision victims)." English summary only. Pages 69–71. DLO-Instituut voor Bos-en Natuuronderzoek, Arnhem, Netherlands. RIN-Rapport 92/2.

— — —.1995. "Bird/wind turbine investigations in Europe." In *Proceedings of National Avian—Wind Power Planning Meeting*, Denver, Colorado, July 1994. RESOLVE Inc., Washington, DC, and LGL Ltd., King City, Ontario. 145 pp.

Winrock International, Global Energy Concepts, and American Wind Energy Association. 2003. *Information About Land Leasing and the Potential for Job Creation Related to Wind Energy Project Development in Mexico*. Report prepared for USAID/Mexico and the State of Oaxaca Secretariat of Industrial and Commercial Development. Arlington, VA, USA.

Wobeser, T., and A. G. Wobeser. 1992. "Carcass disappearance and estimation of mortality in a simulated die-off of small birds." *Journal of Wildlife Diseases* 28: 548–554.

Wolsink, M. 1996. "Dutch Wind Power Policy." *Energy Policy* 24(12). In Damborg, S., *Public Attitudes Towards Wind Power*. Danish Wind Industry Association.

Wolsink, M., and M. Sprengers. 1993. "Wind Turbine Noise: A New Environmental Threat?" University of Amsterdam. In Damborg, S., *Public Attitudes Towards Wind Power*. Danish Wind Industry Association.

Woody, T. 2009. "Judge Halts Wind Farm Over Bats." *New York Times*, December 10, 2009.

World Bank Group. 2008. "Rights and Participation: Citizen Involvement in Projects Supported by the World Bank." *LCR Good Practices Note* 2(1). Washington, DC.

———. 2007. "Environmental, Health, and Safety Guidelines for Wind Energy." In *Environmental, Health, and Safety Guidelines*. Washington, DC.

———. 2004. *Involuntary Resettlement Sourcebook*. Washington, DC.

———. 2003. *Social Analysis Sourcebook: Incorporating Social Dimensions into Bank-Supported Projects*. Social Development Department, Washington, DC.

World Wind Energy Association. *World Wind Energy Report 2010*. World Wind Energy Association.

———. 2010. *World Wind Energy Report 2009*. World Wind Energy Association.

Young, E. P., Jr., W. Erickson, R. E. Good, M. D. Strickland, and G. D. Johnson. 2003. *Avian and bat mortality associated with the initial phase of the Foote Creek Rim wind power project*, Carbon County, Wyoming: November 1998–June 2002. Technical report prepared for SeaWest Energy Corporation and Bureau of Land Management. Western Ecosystems Technology, Inc., Cheyenne, Wyoming, USA.

ECO-AUDIT
Environmental Benefits Statement

The World Bank is committed to preserving endangered forests and natural resources. The Office of the Publisher has chosen to print World Bank Studies and Working Papers on recycled paper with 30 percent postconsumer fiber in accordance with the recommended standards for paper usage set by the Green Press Initiative, a non-profit program supporting publishers in using fiber that is not sourced from endangered forests. For more information, visit www.greenpressinitiative.org.

In 2010, the printing of this book on recycled paper saved the following:
- 11 trees*
- 3 million Btu of total energy
- 1,045 lb. of net greenhouse gases
- 5,035 gal. of waste water
- 306 lb. of solid waste

*40 feet in height and 6–8 inches in diameter

Color Photo Insert

Figure 4.1: Wind Turbine Placement with Regard to Sound Impacts

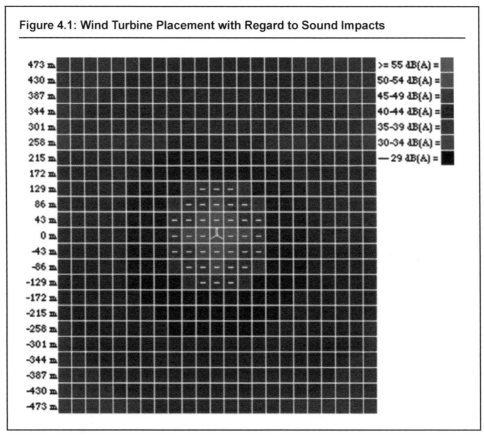

Note: Each square measures 43 by 43 meters, corresponding to one rotor diameter. The bright red areas are the areas with high sound intensity, above 55 dB(A). The dashed areas indicate areas with sound levels above 45 dB(A), which would ideally not be used for residential purposes.

Photo: Roberto G. Aiello

The footprint of land cleared for wind turbine installation includes the area required for the staging of heavy equipment and large turbine parts, as shown here during wind farm construction in Uruguay.

Photo: Carl G. Thelander

The Altamont Pass Wind Resource Area in northern California, USA, has become known for particularly high mortality of Golden Eagles and other raptors.

Photo: Rafael Villegas-Patraca

Black-bellied Whistling Ducks Dendrocygna autumnalis *fly through the La Venta II wind farm located in Mexico's Isthmus of Tehuantepec, a world-class bird migration corridor.*

Photo: Bird Conservation Alliance   Photo: Jim Laybourn

The Greater Sage-Grouse Centrocercus urophasianus *(displaying male, left; typical habitat, right) of the western United States is among the open-country bird species that instinctively stay away from tall structures, including wind turbines and power poles.*

Photo: Ed Arnett, Bat Conservation International

The Eastern Red Bat Lasiurus borealis *is typical of the migratory, tree-roosting bat species that are frequent casualties at some wind farms in North America.*

Photo: Bats and Wind Energy Cooperative (BWEC)

Wind turbines on forested mountain ridges can pose a relatively high risk to bats, as well as specialized ridge-top vegetation.

Photo: Edward Arnett, Bat Conservation International

A turbine at the Casselman, Pennsylvania wind farm, in a feathered state (blades pitched parallel to the wind) during operational curtailment to reduce bat mortality.

Photo: China Wind Power
Wind farms enable most preexisting land uses, such as livestock grazing, to continue in the spaces between each turbine.

Photo: Empresas Públicas de Medellín
The wind-swept areas most suitable for wind power development are sometimes the homelands of indigenous people such as the Wayuu community at Jepirachi, Colombia.

Photo: Roberto Aiello

This historic stone fence, built during the 17th century, runs adjacent to the recently constructed wind turbine facility in the Sierra de Caracoles of Uruguay.

Photo: UTE

Wind turbines (2 MW each), Sierra de Caracoles, Uruguay

Photo: Lygeum
It is much easier to estimate the mortality of large birds, such as this Eurasian Griffon Vulture Gyps fulvus *at a wind farm in Spain, than for small birds or bats that are more difficult to find and are frequently removed by scavenging animals.*